From Protest to Surveillance –
The Political Rationality of Mobile Media

Oliver Leistert

From Protest to Surveillance – The Political Rationality of Mobile Media

Modalities of Neoliberalism

PL ACADEMIC RESEARCH

Bibliographic Information published by the Deutsche Nationalbibliothek

The Deutsche Nationalbibliothek lists this publication in the Deutsche Nationalbibliografie; detailed bibliographic data is available in the internet at http://dnb.d-nb.de.

Printed with financial support of the German Research Foundation (DFG) and the University of Paderborn.

The research partially to this book was sponsored by Central European University Foundation, Budapest (CEUBPF). The theses explained herein are representing the own ideas of the author, but not necessarily reflect the opinion of CEUBPF.

Cover image: A map of Geneva showing digital traces of the mobile phone communication in the city. Copyright Ville de Genève. See http://www.villevivante.ch

Library of Congress Cataloging-in-Publication Data

Leistert, Oliver.
From protest to surveillance : the political rationality of mobile media modalities of neoliberalism / Oliver Leistert.
p. cm.
Includes bibliographical references and index.
ISBN 978-3-631-64313-6
1. Mobile communication systems—Social aspects. 2. Mobile communication systems—Political aspects. 3. Cell phones—Social aspects. 4. Cell phones--Political aspects. 5. Political participation—Technological innovations. 6. —Political aspects. 7. Electronic surveillance—Social aspects. 8. Intelligence service—Political aspects. 9. Intelligence service—Social aspects. I. Title.
HM1206.L45 2013
303.48'3--dc23

2013025879

ISBN 978-3-631-64313-6 (Print)
E-ISBN 978-3-653-03268-0 (E-Book)
DOI 10.3726/ 978-3-653-03268-0

PL Academic Research is an Imprint of Peter Lang GmbH.

Peter Lang – Frankfurt am Main · Bern · Bruxelles · New York · Oxford · Warszawa · Wien

www.peterlang.de

Acknowledgment

Writing is no solo affair. Neither is traveling.
For helping me during the course of this work I cordially thank Hartmut Winkler, Hannelore Bublitz, Theo Röhle, Tobias Conradi, Heike Derwanz, Florian Muhle, Mirna Zeman, Sylvia Kesper-Biermann, Ralf Adelmann, Norbert Otto Eke, Gisela Ecker, Dorothee Meister, Maik Bierwirth, Renate Wieser, Irina Kaldrack, Andre Meister, Kate Coyer, Eva Bognar, Lina Dencik, Kristina Irion, Maxigas, Geert Lovink, Ricardo Dominguez, Sven Opitz, Arne Hintz, Stefania Milan, Devin Theriot-Orr, Amber Voha, Jessi, Abie, Daniel Kahn Gillmor, Nathan Freitas, Elijah Sparrow, Micah Anderson, Rhatto, Elisa, Fernão, Drebs, Diogo, Olivia Janequine, Minerva, Blax, Iokese, Xabier, StartX, Ionnek, Yossarian, Echo, Dinesh, Manasasarovsar, Madhuresh Kumar, Leo Saldhana, Smriti Vohra, Shahid Munir, Uzair and Sundas Hurain, Zaman Khan, Ziyad Faisal, Henrik, Takakao, Takuro, Yasuda Yukihiro, Hirosato Matsuura, Siggi, Hajime, Randy Nobleza, Bianca, Tanja, Ali, Owl, Axel, Jan, Malte, Stefan, Amanda, Elke and Volker Leistert.

Hamburg, 30.4.2013

Contents

1 Introduction

The ubiquity of mobile phones and the ongoing integration of services into mobile devices is a challenging subject to study. This almost global phenomena connects very large parts of populations with a machinic addressable space. It is a medium that connects the literate and the illiterate, those that have no access to the banking systems, the street dwellers, and the children, to name only a few. It appears as if mobile media has no limits. Global mobile phone penetration rates are higher than electricity. Mobile media also brings the internet to new places and milieus. It leapfrogs landline telephony and recently also internet access.

> Most of the people don't know how to read. But every family, even very poor ones, have access to mobile phones. I stayed at a very basic house, a couple with two kids, and they had two mobile phones. And the kids used the mobiles. They didn't have landlines and now they have mobile phones.
>
> Caminaghi / Campinas (Br)

Such a media revolution in an unprecedented invasive mode rearranges many sectors and fields of sociality. It changes subjectivities in many ways and invokes new regimes of powers. The way it restructures, but also reinforces power relations, calls for an analysis from a governmental angle; from an analytics of powers.

A great number of scholars has published a solid body of literature on mobile media. To name only a few: Plant examines the effects of the mobile phone in many different places and cultures (2002). While trying to cover very many topics, and showing both the wide spread use and different practices, the analysis remains brief and reads more like a report. Goggin (2006) approaches the cell phone from the point of view of a device for cultural production, consumption and media convergence. This account is very much focused on the gadget as such. Rich in statistics, documentations and in the style of a textbook, Ling (2004) provides a history of mobile telephony that tries to cover the whole phenomena. Völker combines a genealogy of mobile technology with the history of the idea of virtuality (2010). But the technological stories, which are centered around engineers and inventors, do not in my opinion connect very well to the philosophical reflections about virtuality. There are of course many more monographes, and a fast growing universe of journal articles on mobile media.

The Social Construction of Mobile Media

Anthropologist Raul Pertierra from Manila, who has written extensively about mobile phones in the Philippines (2002, 2006, 2008), argues that

> the mobile, like all other technologies such as the computer and the internet, or earlier ones like the telephone and the radio, are socially constructed and located. Their effects are only the result, even if unpredicted and unintended, of the possibilities that they make available to their human operators. (Pertierra 2006, 16)

While the social construction of technologies certainly is a solid argument, at the same time it lacks explanatory reach: the roll out of mobile media across the globe equals new and conflicting layers of *powers* that structure those human operators in their pursuit of mobile use and subsequent thinking, mentalities and practices. As much as "[cell phones] themselves are embedded in existing social practice" (de Souza e Silva, Sutko, Salis, and de Souza e Silva 2011, 412), the social practice itself is structured, updated and made operational enmeshed in a political rationality.

Thus, the influence that mobile media, by way of an ubiquitous arrangement of numerous different and heterogeneous power vectors, has on subjects, social relations, and society is the subject of this work. It is about the political rationalities of mobile media. The term 'mobile media', that I use, comprises a semiotic-material conglomerate of heterogeneous sources and trajectories; therefore it is not *limited* to the gadget and neither to the data it sends through the networks. And, as already said, not to its social construction. Mobile media comprises the material and the immaterial. Mobile media, in very general terms, here is regarded as an apparatus of security for the production of neo-liberal freedom.

Towards an Analytics of Mobile Media Rule

By pursuing this ambitious and rich theme, the intention is to show that concepts of mobile media need to acknowledge many more facets than commonly investigated. As a political technology it prescribes and describes simultaneously, thus establishing its own rationality and regime. On a very general level, it facilitates the flow of signs, people and goods. Just as James Beniger in his seminal "Control revolution" stated, "*control* encompasses the entire range from absolute control to the weakest and most probabilistic form, that is, any purposive influence on behavior, *however slight*" (Beniger 1986, 8; emphasis in original). Mobile media influences with soft means behavior in many ways. It supplements technologies of control.

When seen under the lens of governmentality, the faster ways of contacting, communicating, exchanging data, and other *productive* constellations that support the flow of things, signs and people can be understood as techniques and technologies of governing. Thus, mobile media is a social machine of circulation. But what are theses virtues and qualities, that mobile media supports, enhances, and strengthens? In what larger picture do they positively resonate? How does mobile media fit into a larger dispositif or arrangement? What are the needs and problems that push mobile media?

The Co-Presence of Signs and Material

Hartmut Winkler, with reference to Beniger, argues strongly in favor of a co-evolution of media with other spheres: "The development of media appears to be *embedded* within the larger context of an economic-technical-social development, which in the first place produces the super structures of modernity that are based on the division of labor"(Winkler 2004, 68; translation O.L.; emphasis in original).[1] In his semiotic-material investigation into the relation of media and economy, Winkler proposes to put the concept of traffic to the foreground, because traffic provides the possibility to "describe goods/commodities and signs albeit their obvious differences in concepts that can be connected" (2004, 93; transla-

1 "Die Entwicklung der Medien erscheint *eingebettet* in den größeren Zusammenhang einer kombiniert wirtschaftlich-technisch-sozialen Entwicklung, die die arbeitsteilig vernetzte Großstrukturen der Moderne überhaupt erst hervorbringt."

tion O.L.).[2] One of the terms that he proposes to investigate amongst others is transmission [*Übertragung*] in the sense of postal delivery. By recourse to Siegert (1993), Winkler asks: "Not only that the state owned postal service demands the addressability of subjects is to be shown but that and why the subjects 'themselves' saw mailbox slots into their mahogany doors" (2004, 108; translation O.L.).[3]

In a similar way, I am interested in exploring why subjects always carry a digital transceiver around, which makes them addressable and integrates them into the circuits of circulations of a political rationality. This echoes Foucault's notion that (neo-)liberal freedom facilitates "the possibility of movement, change of place, and processes of circulation of both people and things" (Foucault 2007, 48-49). Different to Winkler, I am more interested in the political rationality in relation to the emergence of (mobile) media, while I share the assumption that only an approach on the semiotic-material co-presence can convincingly shed light on the problem.

In addition, this work is limited by further specifications: mobile media has reached a point of economical-technical-social integration where it seems problematic for me to formulate a thesis on the subject as a whole. While the angle of an analytics of government provides an entry point to the whole phenomena, the theme here is reduced to two focal points of mobile media: protest and surveillance.

Mobile Protest Media and Mobile Surveillance Media

These focal points are no coincidence. Protests have gained a high degree of enmeshment with mobile media. The mobilization capacity of mobile media, its use within protests for coordination and its easy to use capacity towards media production, such as small films and audio recordings for internet distribution, have been generally acknowledged in the last waves of large protests; maybe most emblematic are the mobile media video documentations from Egypt's uprising. It is

2 "Waren und Zeichen wären, ihren augenfälligen Differenzen zum Trotz, in anschlussfähigen Begriffen neu zu beschreiben."

3 "Zu zeigen eben wäre nicht allein, dass die staatliche Post die Addressierbarkeit der Subjekte verlangt, sondern dass und warum die Subjekte 'selbst' Briefkastenschlitze in ihre Mahagonitüren sägen."

here that mobile media is widely seen as a tool of empowerment.[4] And there is no question that a media technology that is in the hands of everyone, has effects on empowerment and agency alike.

> In my opinion, when we would not have been able to send out SMS, we would not have been able to mobilize the people across the country. We have successfully been able to manage it with SMS across the country.
>
> Mohammad Azhar Siddique / Lahore

But what is gained when one acknowledges this? Many further questions need to be asked: Who gets empowered? What kind of empowerment to which agency is this? What political rationalities are set in the foreground by mobile media? What effects on which subjectivities are the result? To understand which kinds of power relations and which kinds of subjectivities are empowered, is one of the central themes here.

The second leitmotif is surveillance. And again, this focal point is no coincidence. As a political technology mobile media allows many new kinds of surveillances, as well as many old ones, at an incredibly cheap rate. Be it targeted or mass surveillance, be it on the contents of communications or relational about social networks, the issue of mobile media surveillance, be it commercial or by state agencies, is unprecedented. The device on the body, seen from this perspective, is a materialization of pure surveillance media camouflaged; never before was there a digital location tracker that everyone enjoyed carrying around. In the activist universe this has triggered specific fears that lead to radical practice.

> There was a time, when at a social centre they had a sign on the wall and that said 'turn off your phone and take the battery out'. They established this rule that everyone that would go to a meeting had to turn off the phone and take the battery out. The reason being that if you turned off the phone and leave the battery in it, it could be used that people would be able to listen to your conversation through your telephone. It was one of those things that I saw, where I thought: you should be more concerned what you talk in a local bar than taking the battery out of your cellphone.
>
> Anonymous1 / Mexico City

4 See the *mobile active* portal as an example of a hub for mobile media empowerment programs. http://www.mobileactive.org (accessed 4 March 2012).

Both focal points are in a very special relationship, which makes them a premium site of investigation. Mass surveillance schemes like data retention (see section 7) hold two main capacities: on the one hand, they produce a larger picture, or better said, they establish a new knowledge-power axis for governing. The algorithmic production of knowledge about the population's communication behavior entails a whole new set of possible measures for neo-sovereignty. This goes as far as to the production of data derivatives for contemporary pre-emptive modulations of discrimination. The other capacity relates to the first, and advances from there. Once a specific target is calculated, relational surveillance provides access to the social network and automatically shovels more suspects into the profiler's file.

Participants in protests have always been of interest to governments and agencies in many ways; the ongoing habit of police and secret services to collect information about those who go out into the streets is just one. Another one adheres to the problem of governing: one cannot economically govern against the desires of the population. Those articulations of protests need to be channeled into programs of government, they need to be made productive. In this sense, mass surveillance provides abstract information that is available for mining, as if it was statistical data on birth and death rates.

Mobile media serves both: empowerment and surveillance. One might even say: what it gives to one, it just as well gives to the other. But this divide is a problem. It feigns a contradiction. This contradiction is an effect of an oversimplified concept of power, echoing the binaries of "the governing" and "the governed". This is where an analytics of rule is enlightening. At the core, it understands that power is everywhere and foremost inter relational. There can be no societal life without power. The idea is a radical 'democratization' of the power concept: to show that power is effective in different modalities through all fibers of social life; to show that it is a productive force that bridges gaps as it proscribes a continuum of relations.

Ubiquities of Powers and Traces of Rationalities

Foucault, who was the initial figure for such a power model, was often described as a dark thinker; the thinker of the prison (Foucault 1977). A common critique is that he never wrote a book about resistance. An even more common critique is that he reduced the subject to an effect of powers. I cannot disagree more: an analytics of rule can provide a fruitful base to analyze protests and resistances. The paradigmatic shift that makes this possible is to understand government in a very wide sense: that all sociality is interwoven by governmental aspects that connect the subjects to programs of government while at the same time these aspects offer choices and allow the subjects to experience a certain freedom. This shift provides many different ways to look at government and thus understand power. Then, I argue, the rationality, the horizon of truth, that justifies one practice while it delegitimizes others, can be understood and questioned.

To see subjectivity as a primary site of investigation for power relations, takes into account that by pure force no regime can prevail. It is necessary to find those 'interfaces' or points of affirmation that allow a 'system' of devastation and crude social injustice to prevail. If power is seen this way, and one includes into the problem of government the totality of society as multiple productive sites, as a transit passage of governing, a substantial contribution to resistance and protest can be made too, because protests and resistance, by their very own nature, articulate societal issues and demand changes. This makes them a problem of government.

Agency Revisited

There is a third focal point lurking behind the two others – empowerment and surveillance – and in a sense this point fuels this investigation and is its trajectory: how can surveillance mobile media, once it is analyzed and understood, be deprived of its powers? How can communication technology make a difference within the liberal paradox of secured freedom? How can a data collecting assemblage of surveillance technologies be confronted so that it no longer, seamlessly and in a more and more automated fashion invades privacy and social relations with a single click from an operator's desk? The easy, but cynical answer is, by switching off the mobiles. But there is a point here: while it has become increasingly problematic to partake in social life without a mobile phone, because mobile media is invested with great powers, the question is why are there no privacy enhancing technologies in wide use?

If one divides surveillance into surveillance of contents and of relations, the former can be encountered with privacy enhancing technologies, while the latter is situated in a technical necessity, that of meta data or transactional data. Therefore, the retention of transactional data is an operation that benefits from the neo-liberal scheme of mobile media. At the same time, it is an expression of the paradox of liberal rule, which excessively tries to control what it cannot control. Because meta data cannot be avoided in mobile communication, it does not help much to use privacy enhancing technologies here. Only in self-operated infrastructures, this I try to show in part three on bottom-up strategies, can the problem of meta-data be faced. And here the interviewees come into play.

In this work the discussion of mobile media as, a facet of governmental rule, was informed by 50 experts who I interviewed in 2009.[5] These interviewees actually guide the discussion to large parts as they provide the insights to practice. Be it mobile protest media or the effects of surveillance, it is their encounters which have informed the analysis and which made sure this research is based not only on theoretical considerations but grounded in narrated practice. The interviewees are reporting about their experiences in protests, how they use mobile media and what surveillance means to them.

But one has to be cautious here: no matter how much one refers to the interviews as expert interviews, that are analyzed as discourse, they remain a problematic hinge of knowledge transposition into an analytics of rule. On the other hand, the gains are high: what they show is a discourse about protests and surveillance that has a global reach.

Although this work it not about the goals, detailed interests and actions of the interviewee's groups and movements as such, I still felt a need to contextualize them briefly. This is done in appendix C where, as much as possible, the interviewees themselves explain their activities. Ethical considerations can be found in appendix A. The problems treated there consider academic knowledge production about activists that engage in fights for social justice. It provides information about measures taken to protect the sources, and the problem of language and dissemination of this work.

Mobile Protest Media Case Study and Mobile Surveillance Media Case Study

There are two case studies present in this work that propose an understanding of mobile media from two very different ends as a political technology: The first is about the lawyers' movement in Pakistan, which took place in 2007 and 2008. My

5 This round number is a coincidence.

interviewees here provide detailed accounts of the first mobile media supported civil society movement in the history of Pakistan. The case provides many good arguments for a problematization of mobile media empowerment, as it shows how a governmental subjectivity is facilitated by mobile media. The state of surveillance during the uprising of lawyers against the Musharraf regime was not able to cope with the unrest: it was not yet upgraded to deal with the governmental powers of mobile media - a historical situation that very vividly describes the dialectical relationship between these two facets of mobile media.

The second case study rests on court documents from Germany, which provide accounts and insights into the practice of surveillance in a Western, liberal democratic regime. It describes mobile media in its neo-sovereign mode: how suspects are *produced* by the agencies themselves with technological means. After years of an unleashed and total surveillance operation, without any outcome that would be relevant enough to open an orderly case in the courts, the agencies continued to invade the lives of the Berlin activists. Shutting off the mobile here triggered suspicion to the point of continued observations. The very intense use of sending silent SMS to the suspects' phones provides an illustrative example about the locational tracking function of mobile media.

Global Governmentality via Mobile Media

Many of the interviews are conducted in countries other than the West. One may ask about the applicability of an analytics of rule in countries as different as Pakistan, the US or Japan. Or wonder how a concept, which was developed vis-à-vis to Western societies is applicable to non-Western societies.

I discuss this innovative claim in detail because, for the theme of mobile media as a political technology, one can make very good arguments for an applicability beyond 'the West'. My argument is that (neo-)liberal rule is not bound to institutionalized democracy and not to 'the West' as a geographical region. Much more, *where there is mobile media, there is some sort of liberal rule.* Mobile media can be seen as a product of a globalized communication regime, which facilitates and 'transports' many mental states and habits once it is rolled out on a massive scale. While the degrees and outcomes may vary, in general it enables to jabber free from the state and from traditional ties. This stregnthens a specific rationality: that of the individual. It is mobile media that exports, with its own reductive modifications and means, the Western concept of the individual globally.

Architecture of this Work

To summarize: this work is constructed in three parts. Part one *Play of Freedom* introduces the field of governmental studies in relation to mobile media. While Michel Foucault introduced the concept of governmentality in his late lectures, by now many other scholars have contributed to the field.[6] As much as Foucault's texts, it is their texts as well, which informed the outline of the field and other chapters.

Part one also discusses the conflict-prone relation of empirical research and governmental studies, and the applicability of governmental studies and generally of an analytics of rule onto non-Western societies. What follows is an attempt to come to terms with resistance, empowerment and protests within the analytical framework. Finally, part one provides the *mobile protest media* case study about the lawyers' movement in Pakistan from where it finally develops a frame of distributed action via mobile media.

Part two *The Liberal Paradox* starts with a summarization of surveillance studies and concepts of surveillance. It further discusses in depth what I call *algorithmic neo-sovereign rule*, which can be seen as an illiberal trajectory within the liberal paradox. Under discussion is data retention and databases on specific suspects. Blocking of mobile communications and the mandatory registration of SIM cards are further examples where liberal rule becomes illiberalized. The closing part then comprises the case study of the §129a investigations against activists in Germany, which provides material evidence about a mega surveillance practice in a 'democratic' country.

The third, final, and shortest part *Bottom-Up Strategies*, navigates the terrain of how mobile communications technologies can be freed from illiberality. Concepts from the technological realm are presented which seek out central weak points that allow the relatively cheap operations of the contemporary surveillance assemblage.

---

6 To name only few: Nicolas Rose, Thomas Osborne, Andrew Barry, Thomas Lemke, Susanne Krasmann, Hannelore Bublitz, and most importantly Mitchell Dean, whose "Governmentality: Power and Rule in Modern Society" is used extensively here for its clarity and almost didactical introduction to an analysis of government.

As closing remarks, instead of lengthy recapitulations, I formulate a couple of closing arguments developed during the course of this work. They are not a summary and need to be seen in relation to the former chapters.

As a whole this work tries to make a contribution to the question of *how* mobile media could have become so extremely successful. The sites of contestation – protests and surveillance – that are under consideration, point to the limits of mobile media. It is here that the trajectories of mobile media crystallize.

Part One: Play of Freedom

2 Governmentality

Introduced by Michel Foucault, in his late lectures (see 2007 and 2008)[1], the study of governmentality aims at understanding how subjects are constituted as governable and how governing produces subjectivities whose rationalities are in a productive relation to its surrounding societal settings. Within its genealogical account it aims to understand *how* it has been possible that economy has been constituted as an object of government, and thus how as a practice of economical governing, how such a *rationality* has been produced.

This trajectory is a shift in Foucault's theories of power and it establishes a third pillar, that of governing as *management* of populations. While he very intensely analyzed what he calls sovereign power and disciplinary power, he now introduced the concepts of a third type of power with this "rather awkward neologism" (Miller and Rose 2008, 15); *governmentality*.

Foucault differentiates a triangle of power with different functional modalities and problematizations:

> Let's say then that sovereignty capitalizes a territory, raising the major problem of the seat of government, whereas discipline structures a space and addresses the essential problem of a hierarchical and functional distribution of elements, and security will try to plan a milieu in terms of events or series of events or possible elements, of series that will have to be regulated within a multivalent and transformable framework. (Foucault 2007, 20-21)

I will come back to the concept of security, as it is the most important element here. For now, I want to emphasize that these 'power settings' consist of strategic

---

1 Other important texts in this matter are Foucault 1997a, 1997b, 1997c, 2000f, 2000g, and in part his lectures 1975-76: 2003.

and tactical techniques, and technologies of governing, in a different and much wider sense than political science traditionally understands *government*:

> By 'governmentality' I understand the ensemble formed by institutions, procedures, analyses and reflections, calculations, and tactics that allow the exercise of this very specific, albeit very complex, power that has the population as its target, political economy as its major form of knowledge, and apparatuses of security as its essential technical instrument. (Foucault 2007, 108)

Foucault identified a shift, or transgression, in government between the 17th and 19th centuries, which led to new "art of government" that had new core issues. The new trajectory articulated an answer to the old problem of the state: how can the state survive? This also means that the aim of governing changed from mere sovereignty, which at its center is answering problems of how a sovereign is related to a territory and disciplinary power, which targets the production of "docile bodies", to governmentality, one that targets the population via statistics:

> The final objective is the population. The population is pertinent as the objective, and individuals, the series of individuals, are no longer pertinent as the objective, but simply as the instrument, relay, or condition for obtaining something at the level of population. (2007, 42)

This paradigmatic shift is expressed in that governing no longer understands a series of individuals as the target of government, but an abstraction: the population.

Economical reasoning is finding its way into the "art of government" and shifts the rationality of government. This "funding act of economic thought and economic analysis" sought a solution to the problem of scarcity, as scarcity leads to revolt: "freedom of commerce and of the circulation of grain began to be laid down as the fundamental principle of economic government" (2007, 33). The early modern state's "essential issue of government will be the introduction of economy into political practice" (2007, 95). No longer can sovereign and discipline powers alone respond to the needs of a population since its individuals are framed as free citizens with rights. The predatory style of sovereign power as invoked in the classical age could not handle the need to react on issues of scarcity of food. It invoked unsuccessful measures as it was blind towards the population as a productive force.

Governmental regimes, that co-emerged with the modern state, are then historically related to bourgeois revolutions. Presently, political economy has become a structuring pillar of government globally, and even within the most rigid regimes.

In relating a population's "natural" behavior, such as trade and exchange, to a technique of governing, the emerging modern state was much more effective and less costly then prior modes of rule, whose exercise of power towards its subjects was primarily structured through exertion of force or violence, i.e. one of the most common logics of rule was concerned with taking, not granting (life, food, money).

Thus, for the first time in the history of governing, economic considerations became the guiding principles. This could only be established with a whole new perspectives on what ruling is about and what it encompasses. Foucault delivers a deep description of the historical processes that brought to light what today is seen as liberal rule. This genealogy is not of particular interest here, so I touched it rather briefly.

Nikolas Rose and Peter Miller point out the wide range of what counts as government, as it is the:

> historically constituted matrix within which are articulated all those dreams, schemes, strategies and maneuvers of authorities that seek to shape the beliefs and conduct of others in desired directions by acting upon their will, their circumstances or their environment. (Miller and Rose 2008, 54)

My claim is that this concept fits in understanding mobile media, mobile phone dissemination, and the production of the corresponding subjectivity, which has a need of "unlimited" communication because it is enmeshed in such a matrix.

> Mobile phones are a very basic working tool, because our system here works very much via information exchange. Before mobile phones it was very hard: we went many times and waited for a call or go and communicate with people personnally. We were using phone cards. But sometimes people were late and couldn't tell that everything was alright, because they he had no phone card. Then some people would go and look for that person. Now the mobile phones solve 80 percent of our external communication needs
>
> Bispo and Christoph / São Paulo

Mobile media, to my understanding, is a promising candidate of political technologies to discuss what "is precisely the nature and limits of 'the political', the political as itself a transactional space, a historically variable zone of rationalization and division" (Barry, Osborne, and Rose 1996, 7).

> There is an impressive phenomena in Mexico: the explosion of the use of mobile phones. In a few years it became much more than landline users. The social organisations are not an exception. Very fast it became from something that only the business people used to a technology that is used massively. I perceived something similar with laptop computers, there are more offers to get one, there are credits.
>
> Enrique / Mexico City

To affect will, circumstances, and environments are techniques that comprise what government in its modern design is aiming at. In the first moment, to 'act upon will' seems to contradict the notion of the autonomous subject and the idea of free will.

Political Technologies: Pastoral Power

But how can the modern governing work in this way? What existing techniques were available that would allow individuals to be conducted *and* be self-conducted? On the individual level, the historical cultural quality that was made productive came from Christian tradition.

> Christianity had in large part organized power's hold on the ordinary preoccupations of life: an obligation to run the minuscule everyday world regularly through the mill of language, revealing the common faults, the imperceptible failings even, and down to the murky interplay of thoughts, intentions, and desires. (Foucault 2000a, 166)

Originally, it was Christian pastoral power that made subjects susceptible to guidance. With the emergence of the modern subject the ability to guide the self-conduct of subjects has been adopted by the modern state in a secular form (Dean 2010, 90-101; see also Bublitz 2010a, 57-77).

The name "pastoral power" seems irritating as this coincides with the emergence of secular rule and the secular state, but Foucault argues that "the modern

Western state has integrated into a new political shape an old power technique that originated in Christian institutions" (Foucault 2000f, 332). This "ritual of confession" as an "individualizing power" (Foucault 2000g, 300) was outreached by an administrative power, "a recording mechanism instead of a pardoning mechanism" (Foucault 2000a, 166). This cultural technique of confession served different aims within secular powers. One was to bring "the quotidian into discourse, to survey the tiny universe of irregularities and unimportant disorders" (Foucault 2000a, 166) and further to accumulate, make productive and transform these former traceless uttering into a whole new body of knowledge.

Act Upon Will

Pastoral power provides the qualities to "channel" subjectivity because it works on the level of mentality: it looks after the individual, it brings to light and into discourse the inner thoughts and mentalities of the individual. It harvests a knowledge of the conscience and performs an ability to direct it. This is where the famous "conduct of conduct" gets under way, which has found its historical programmatic fulfillment in neo-liberal government (Bröckling 2007). This mode of governing conducts subjects as if they conducted themselves. Hence, governing upgraded to the 'free will' of the modern subject. Pastoral power emerged when the modern subject was invested with rights and autonomy. The modern state, in its political and philosophical conception, at the same time produced a power independent from 'the Political' (with a capital P). Pastoral power operates vis-à-vis a "field of possibilities in which the behavior of active subjects is able to inscribe itself" as "a set of actions on possible actions" (Foucault 2000f, 341).

I argue that mobile media seamlessly integrates into this form of power. Mobile media encourages users to chat. It expands the plane of pastoral power to a technology over distance, which can inspect and nurture the quotidian and intimate at any given place and time. Mobile media answers to the problem of mobility, and it updates pastoral power to this new problem.

Pastoral power is thus one key ingredient of governmental power: it conducts and its conduct is rationalized as freedom, as it conducts with offers and possibilities much more than with constraints and prohibition. Techniques of confession are part of this complex set, such as the Freudian analysis or self-help-groups in feminism (Illouz 2008, 97-103). Recent television formats are part of this set of techniques; empowering the subject to discuss its inner state in public to extend that subjectivity is a product of such self-presentations, a process that can be called post-disciplinary subjectivation (Bublitz 2010a).

In contrast to previous modalities of governing, now the subject was to be convinced of acting within a specific desired realm of addressability and rationality, where it could be held responsible as an individual invested with rights. In former modalities of governing the subject was to be disciplined at its outer layers; to make the body work, one might say. Governing had not developed political techniques to questions of mentality. The field of mentalities and conscience formerly had exclusively belonged to religion and the Saviour of the soul.

What changed is that governing started to rationalize this resource as it co-functions with the goal of economic governing: let the subjects care for themselves. The way in which freedom is organized and laid out by modern governing has many facets, but maybe most importantly "governing involves not just the ordering of activities and processes. Governing operates through subjects" (Miller and Rose 2008, 42). This can only be be achieved if the subjects are addressed as individuals with a formal autonomy and rights. Government becomes a personalized matter now which enrolls "individuals as allies in the pursuit of political, economic and social objectives." Government programs thus become a "part of the 'self-steering' mechanisms of individuals. Hence 'free' individuals and 'private' spaces can be 'ruled' without breaching their formal autonomy" (Miller and Rose 2008, 42). Pastoral power does not contradict a concept of formal autonomy, as it directs the formal autonomous subject and its decisions in a non-subjugating, soft and rationalized manner. A convinced subject acts as a free subject.

Freedom as Source of Governing and as the Correlative of Security

Enrolling individuals as allies in the pursuit of political, economic and social objectives needs to rely on a broader and much more extended net of operational entities and discourses. Miller and Rose are proposing different problematizations of political rationalities "in which politicians, intellectuals, philosophers, medics, military strategists, feminists and philanthropists have measured the real against the ideal and found it wanting", which *translate* into programmatic work

> of designs put forward by philosophers, political economists, physiocrats and philanthropists, government reports, committees of in-

> quiry, White Papers, proposals and counterproposals by organizations of business, labor, finance, charities and professionals that seek to configure specific locales and relations in ways thought desirable. (Miller and Rose 2008, 61)

The rationality of political economy is translated into programmatic modalities. Foucault characterized governmental power thus as "centrifugal", because the elements government seek to render into governing are, "in principle unlimited" (Miller and Rose 2008, 63).

One can call the resulting heterogeneous ensembles for governmental management *assemblages*. They entail

> the routines of bureaucracy; the technologies of notation, recording, programs, knowledge and expertise that compose a field to be governed and invest it with purposes and objectives; the ways of seeing and representing embedded in practices of government; and the different agencies with various capacities that the practice of government require, elicit, form and reform. (Dean 2010, 37)

The advantage in describing governmental rule assemblages is that any reference to this rule as a single command structure or other pure instrumental orders of rule is excluded. An assemblage is a rhizomatic structure. It comprises links to sovereign or disciplinary rule, but not on a privileged level. In the detailed discussion about data retention I return to the concept of the assemblages how it is used by Haggerty and Ericson (2000) to understand the field of surveillance.

Economy, Freedom and Security

Foucault's earlier writings, most prominently on discipline and punishment, did not theorize much on the question of how a regime can remain in power productively and in a sustaining way. To suppress the subjugated is expensive and it cannot be done in a very fine granular way at the individual level. Further, it has only short term effects compared to an organization of government in such a way that the subjugated deliberately act in concord with rule. With the turn to governmental management the key question becomes the "economic relation between the cost of repression and the cost of delinquency" (Foucault 2007, 9). Thus, Foucault has offered a reading of government that aims explicitly at governing "through a particular register, that of the *economy*" (Dean 2010, 29; emphasis in original). The freedom produced in such a rule "is nothing else but the correlative of the deployment of apparatuses of security"(Foucault 2007, 48). Therefore, as freedom is about the flow of things, signs and people, programs have to be deployed, like mobile media, that connect to the technologies of the self.

As Thomas Lemke writes,

> Foucault corrects the findings of the earlier studies in which he investigated subjectivity primarily with a view to "docile bodies" and had too strongly stressed processes of discipline. Now the notion of government is used to investigate the relations between technologies of the self and technologies of domination (2002, 4).

Economic freedom then does not refer to abstract aims, but to a civic mobilization at a large scale: mobilizing the whole population as agents of government translats abstract political economy into "a formula of rule, a resource for government" (Barry et al. 1996, 8). The population follows its 'own interests' without fearing sanctions in this pursuit.

Governing from letters to mobiles

Bernhard Siegert has discussed the establishment of the postal system for citizens (Siegert 1993). By way of monopolizing postal correspondence, he argues, the state has turned away from a Machiavellian view on population to a bio-political view, which regards population as a productive entity: the postal system thus had to serve the benefit of the population; population became its reason for existence. "A goal of the postal system is to enmesh people in discourse" (Siegert 1993, 62; translation O.L.), while at the same time, the ubiquity of the postal system acted as an identification scheme: "Identity is not longer a question of biographical depth, but a question of addressability" (Siegert 1993, 126; translation O.L.).

The postal system transforms pastoral power into a source of government and, just like mobile media, demonstrates how power is dissipative and exhausts voluntariness of citizen cooperation. Governmental rule organizes a milieu for circulation, that of letters, signs, and, according to Siegert, literature. The circulation of letters ensures an existential precondition of freedom as it functions as an apparatus of security to organize this milieu.

When subjectivities are analyzed as products of this relation to freedom at the core of government, then liberalism is not "an absence of government, or of lessening of political concern with the conduct of conduct" (Barry et al. 1996, 9) as neo-liberal ideology proclaims. On the contrary: government has nested itself into the tiniest strains of intelligibility, sociality, and affects to produce specific actions. These means that allow governing to nest itself within these tiny bits are what Foucault calls the apparatuses of security.

They are the products of programs of government mentioned by Miller and Rose. The apparatuses of security translate reality as they produces milieus. But as they deal with a complex and heterogeneous material, so the say the 'dirty reality', the precise outcome of their operations is not immediately known nor foreseeable. Statistics deliver only a certain kind of knowledge about the reality. Calculation thus plays an integral role in governmental management. The space of security where economic freedom is produced "refers to the temporal and the uncertain", and this milieu is what is needed "to account for action at a distance of one body on another. It is therefore the medium of an action and the element in which it circulates" (Foucault 2007, 20-21). Freedom, in this understanding, is not an antagonism to security but its correlative. Freedom is the intended effect of a milieu production, that of the milieu of security. Free subjects are "individual or collective subjects who are faced with a field of possibilities in which several kinds of conduct, several ways of reacting and modes of behavior are available." In this design "face-to-face confrontation of power and freedom" is impossible as "mutually exclusive facts" (Foucault 2000f, 342). The freedom of choices arranges a plane for power relations to act upon, with and without friction.

Freedom in this concept is not transcendental or metaphysical, but in very concrete terms the ability to act upon choices, whose probabilities are laid down by governmental techniques and the apparatuses of security.

Of course, governing is not only operating in the modality of probabilities to regulate reality, as it is "possibly making use of some instruments of prescription and prohibition" (Foucault 2007, 47) (read: discipline and sovereign power). Security includes any means necessary to ensure the goal of the production of freedom, although primarily security is not an enforcing, suppressing technique. Security has to come to grips with desires as "every individual acts out of desire. One can do nothing against desire" (Foucault 2007, 74). Economic government, and thus the survival of the state, depends upon the viability of "how to say yes to this desire" (Foucault 2007, 73).

Population and Society

This leads to an unprecedented need for knowledge about the governed objects / subjects, which, as they formulate their desires and bring them into discourse, become the subjects of government.

The management of knowledges about the population is informed by "a set of elements in which we can note constants and regularities even in accidents, in which we can identify the universal of desire regularly producing the benefit of all" (Foucault 2007, 74). From the production of statistics, to the detailed and special interest in subgroups and subsets of the population, the new managerial art goes much deeper into people's lives and affairs than any governing ever before ever has: "managing the population means managing it in depth, in all its fine points and details" (Foucault 2007, 107), to the extent that statistical knowledge can provide.

A Nominalist Perspective on the State

When Foucault says that "governmentalization of the state has nonetheless been what has allowed the state to survive" (Foucault 2007, 109), his emphasis on the survival of the state is more than a comment on state centered thinking (which itself can be understood as a product of governmentality). It derives from the notion that from the point of modern governing "one can be nominalistic about the state: it has no essential necessity of functionality" (Miller and Rose 2008, 56), or otherwise, one can call "this a radical political constructivism, a radical constructivist perspective on the state" (Saar 2010, 39). This is a provocative assessment, contrary to the established modern tradition to root political theory in concepts of sovereignty that relates to feudal or absolutist phases (see Krasmann and Volkmer 2007, 10-12). For the understanding of the historicity and political rationality of rule, this is a logical step: the state has no essence, no non-historical core and lacks core mechanisms. Its sedimentations are reflections of historical problems. Thus, it is a "way of dividing a 'political sphere', with its particular characteristic of rule, from other 'non-political spheres' to which it must be related" (Miller and Rose 2008, 56). And this division is in flux.

Governmentalization of the state essentially led to the emergence of civil society. With the emergence of this 'counterweight', the state has naturalized and immunized itself by artificially dividing government spheres: that of the state and that of civil society.

Government-divided spheres into Political / state and non-political / non-state spheres, albeit still both remain terrains of governing. As government "is at the same time both external and internal to the state", it is potentially limitless and total. It decides, "what is public and what private, what is and is not within the state's competence" (Foucault 2007, 109). But while this government identifies and manages realms and fields outside of 'Politics' it tries to do so "without destroying its existence and its autonomy" (Miller and Rose 2008, 60).

This domain is what today we refer to as *civil society*. Protests and resistance emerge from this civil society's initiatives. Civil society can be understood as the counterweight of the state, as in liberalism (and in general in the modern state), "political rule was given the task of shaping and nurturing that very civil society that was to provide its counterweight and limit" (Miller and Rose 2008, 59).

Mobile media's success is a success based on the penetration of civil society with this political technology.

> The mobile is increasing, Bangladesh is connecting, there are 50 million people now, there is also a sort of GPRS connection. Maybe there are 6 companies in the country. It is increasing, mostly young people, so 50 million does not mean 50 million people, but SIM cards. Using them reduces the cost to communicate with each other, so in general, activists are using email or the mobile so that groups come together, to call "let's have a meeting in the evening". It helps to reduce cost of writing letters. They come much closer in one time frame. Suppose there is a rally, so you text it to people to come and join. It is very useful
>
> Swapan / Dhaka

Civil society is once a necessity in functional and economic terms of governing, and it is once the state's reason to survive as the state acts as its counterweight. Together, they form a functional, reciprocal entity; its devision is constantly being renegotiated. Civil society is "both distinct from political intervention and yet potentially alignable with political aspirations" (Barry et al. 1996, 9). It can function as a counterweight to the state because here "power is confronted on the one hand, with subjects equipped with rights that *must not* be interdicted by government." In economic terms civil society fulfills a function within government, as government "addresses a realm of processes that it *cannot* govern by the exercise of sovereign will because it lacks the requisite knowledge and capacities" (both Miller and Rose 2008, 59; emphasis in original). Civil society is the answer to the problem of how to govern economically as it allows the realization of processes that are not the realm of sovereign rule. Civil society thus emblematically stands for what liberal rule has produced: an new zone of rule with its own regularities, that acts as a correlative to sovereign rule and at the same time as the sovereign's (the state's) raison d'être. Indeed, "the objects, instruments and tasks of government must be reformulated with reference to this domain of civil society with the aim of promoting its maximal functioning" (Miller and Rose 2008, 59). From this it follows that protests originate from civil society and address the malfunctioning of government. Hereby protests themselves turn into an integral part of government.

> [T]he idea of an opposition between civil society and the state was formulated in a given context in response to a precise intention: some liberal economists proposed it at the end of the eighteenth century to limit the sphere of action of the state, civil society being conceived of as the locus of an autonomous economic process. This was a quasi-polemical concept, opposed to administrative options of states of that era, so that a certain liberalism could flourish. [...] the reference to this antagonistic pair is never exempt from a sort of Manicheism, afflicting the notion of the state with a pejorative connotation at the same time as it idealizes society as something good, lively, and warm. (Foucault 2000e, 371-372)

The administrative state and civil society have a relationship of mutual dependency, where civil society serves as an identification figure in contrast to the state. There exists a "mutual interdependence of power, freedom and resistance" (Death 2010, 239) in governmental management which is essential for the survival of the state. By addressing the state and demanding participation civil society re-enacts the state, appeals the state, calls for the state. It produces its raison d'être by this invocation.

Governing then is the art of integration and production of performative answers in regard to articulations and aspirations of the population: practices aiming at channeling these aspirations towards a compatibility of economic production, circulation of any sort, while the effect is maintaining the state. Governing is concerned with the survival of the state, an aim that is circular and infinite.

Mobile media as a political technology of civil society, as a governmental technology, is a heterogeneous ensemble that adapts and is flexible, and has produced specific practices, by which it can be identified. These practices have to be rationalized to become real. Bröckling et al. summarize this point nicely:

> It is rationalities and practices of government that generate subjects in the first place, subjectivating by invoking and legitimizing certain images of the self while excluding others. People are thus addressed, for example, as citizens aware of their rights, as political activists concerned about the societal future, or as artists who realize themselves through their creativity; and they also articulate themselves as citizens, activists, artists, and so forth. (Bröckling, Krasmann, and Lemke 2010, 13)

Invoking the image of a productive self, whose communication needs are structured by the paradigm of circulation, mobile media is a strong facilitator of subjectivity production and rationalized practices. I will return to this later.

Sovereignty and Discipline in the Age of Governmentality

> We do have a legislation that is not much in any way with the practice. We do have a legislation that requires, when my phone is tapped, a permission from a particular level of authority by the police for doing so. The authority from whom the permission has to be sought, for political reasons their phones were getting tapped and they never agreed on this. The police is very privileged with the communications department in India. You just need a normal constable in India. Nothing has to be given in writing, nothing has to be given as evidence, no reason has to be quoted for doing that. The police access to the communication department has been just disastrous in that they really end up not following any legislation. If you ask the station head officer: do you know that there is a legislation which prohibits you from tapping my phone? He would laugh at you.
>
> Vijayan / Delhi

The power modulations of sovereignty and discipline (prohibit and punish) are being rearranged, eventually in surprising ways, to introduce management and governmentality into the modern state, as they have to be arranged within a heterogeneous set of managerial apparatuses. Subjects now respond differently to such prohibitiona and punishment once they are addressed as governmentizable. Therefore any government runs high risks when it invokes its 'hard' registers too often or too contingently.

> The mobile phone network was absolutely working during the 2006-7 protests. Without interruptions. That is really strange, because Oaxaca was a strategic point of resurgency and all this shit since 2006. In one minute they can shut down all the mobile phones, but they didn't. I don't know why. Maybe capitalism is bigger than we think. Here in Mexico, the boss of the cell phone networks is one of the richest in Mexico. Maybe in situations like this, surveillance and network are the same. This all is part of the contradiction
>
> Blax / Oaxaca (Mx)

Prohibition and punishment must arrange themselves within the established transit reality of society as "the emergence of a new field for producing effects of power: the new, self-regulating field of the social" (Scott 1995, 203) resonates with all power modalities. While discipline produces "docile bodies", governmental management produces mindsets, rationalities and performative settings that relate the subjects to the state: it can no longer appear as an exteriority of incalculability and fear, but has to rule in accordance with the population's needs and desires.[2]

Government Programs: Techniques, Practices, Failures and Risks

> More and more people in the movements are using 3g cell phones and MST has been implementing lately IP-phones for those activists that travel. So, anywhere in the world they can talk for free.
>
> MST Workers / São Paulo

Techniques and technologies of governing are not reifications, materializations of technologies, or gadgets of some sort. Governmental studies are not technocentric. Much more, "to analyse mentalities of government is to analyse thought made practical and technical" (Dean 2010, 27). This means that the

2 I discuss the emergence of a neo-sovereignty that operates within this paradox of liberal governing in chapter 7.

> concept of technology here includes technical artifacts, strategies of social engineering, and technologies of the self; it refers to both arrangements of machines, medial networks, recording and visualization systems, and so forth, and to a range of procedural devices through which individuals and collectives shape the behavior of each other or themselves. (Bröckling et al. 2010, 12)

Thomas Lemke points out that the

> analysis of government operates with a concept of technology that includes not only material but also symbolic devices. It follows that discourses, narratives and regimes of representation are not reduced to pure semiotic propositions; instead, they are regarded as performative practices. (2007, 9)

It is the semiotic-material ensemble, which engages in actions, practice and channels behavior. Thus, a governmental technique is successful when a practice is commonly established, in wide use and performative.

> I just received a phone call for negotiations, but I don't finish it by cell phone, because human to human contact is much richer. But this way the system works today, you need a phone. Technology is created and so you must engage. Today's technologies are bad for the environment, and the capitalist system and consumer society creates an environment where you need those technologies
>
> Bispo and Christoph / São Paulo

Mobile media is such a conglomeration of technologies, bodies and symbols, which enable performative practices with different effects. It can be understood as a channeled proposal of an unlimited demand for performativity that answers its own circuits, as it prescribes and describes conduct at the same time.

Nikolas Rose highlights the difference of this conduct to other forms of human reflection. "No doubts throughout the ages humans have reflected upon the conduct of themselves and others, but such thought becomes governmental to the extent that it seeks to render itself technical, to insert into the world by 'realizing' itself as a *practice*" (1996, 41; emphasis in original). This is an established setting of rules, conduct, and norms. Practice in the governmental sense is always a historical product of a heterogeneous ensemble that together sets up this practice, but is never ultimately fixed. Practice as part of reality means it is always comprised of a larger operational set of functionalities to deal with the undetermined complexities of reality.

> This is a kind of people's assessment. With social movement people, for them the social movement is their goal, so first social movement, and then: how to use mobile media for that. But general public, for them, mobile media exists from the beginning in their daily life. They are immersed in mobile media. When they felt that they want to participate in a social movement, they just use what they are used to use
>
> Two Korean Activists

Regimes of practice naturalize on the one hand, and immanently *include* alternatives, updates and actualizations towards specific problems on the other. They need to renew themselves constantly to remain operational. They "possess a logic that is irreducible to the explicit intentions of any one actor but yet evinces an orientation toward a particular matrix of ends and purposes." (Dean 2010, 32)

Due to the systemic limits of governmental knowledge, governmental practices are continuously in the process of iteration and updating as it aims to work productively with the population. Direct intervention into sub-groups is costly and to be deferred as long as possible. Political economy factors decide on the viability of a regime of practice. This effect mobile media practice everywhere and sets the horizon of possibilites.

> The japanese mobile media costs are high, so even when you like it, it is difficult to use. If I had more money, there would be a lot of things to do with it.
>
> Yasuda / Tokyo

This necessary constant adaption of governmental practice, which includes failures to certain limits, is systemic in so far as it works vis-à-vis an abstract: the statistical knowledge about the population on which it decides and reflects. But "solutions for one programme tend to be problems for another" (Miller and Rose 2008, 71). By nature, government risks malfunctioning or producing unintended effects. Programs of government are "confronted with forces removed from their access or blocking it, deflecting it, or neutralizing it, these programs also consistently go astray" (Bröckling et al. 2010, 11). Government cannot adjust to the particularities on all levels of scale and in real time. "Hence the description and prescription always involves elusion in that knowledge of government is always also an erring, inadequate, or failing knowledge." (2010, 11)

For this reason programs of governing are invested with a calculation of their own risks. The reality has its own resistances towards programs as it appears "to escape those bodies of knowledge that inform governmental programs, refusing to respond according to the programmatic logic that seeks to govern" (Miller and Rose 2008, 71). But resistance from reality and inadequacies of programs are part of the calculus of governmental management. Failure is a feature, it is an intrinsic, systemic element of governing.

> Unplanned outcomes emerge from the intersection from one technology with another, or from the unexpected consequence of putting a technique to work. Contrariwise, techniques invented for one purpose may find their governmental role for another, and the unplanned conjunction of techniques and conditions arising from very different aspirations may allow something to work without or despite its explicit rationale (2008, 71)

For the conglomerate of protest, mobile media and surveillance, which comprises, evokes, and updates a large set of governmental programs, it is likely to have failing pieces of programs and new outcomes that are unexpected or interfer with an outlined rationale. I will return to this in the sections that discuss technologies of securitization and surveillance.

There is no master plan governmental techniques can follow, but they prescribe and channel according to what they presume to find in reality. Their "programmes presuppose that the real is programmable, that it is a domain subject to certain determinants, rules, norms and processes that can be acted upon and improved by authorities" (2008, 63). The outcomes of the programs operate themselves in the plain and open, struggling to form a milieu of security. Governmental practice thus produces a truth by acting in the real, a truth that only exists if it performs productively in reality. Truths need vigor, agency. Regimes of practice "can be identified whenever there exists a relatively stable field of correlation of visibilities, mentalities, technologies and agencies, such that they constitute a kind of taken-for-granted point of reference for any form of problematization" (Dean 2010, 37). The conduct of mobile media is such a taken-for-granted point of reference. As mobile media connects the individual technologically to regimes of communications, it emblematically follows a trajectory of integration into the circuit of addressability and thus governing. One might understand it as a programmatic empowerment of the subject to governmental rule.

> One problem in occupations we always have is the batteries and how to charge them. We don't have electrical power. That is one of the reasons why we try to make contacts with the people around the occupation: to be able to charge the phones. Usually within 2 weeks we get a clandestine powerline.
>
> Fernão / São Paulo

If "powers are stabilized in lasting networks only to the extent that the mechanisms of enrollment are materialized in various more or less persistent forms" (Miller and Rose 2008, 65), and thus these lasting forms act as stabilized channels for established regimes of practice, their modification, rejection, or affirmation is a trigger and strategic point for societal change, as "'Power' is the outcome of the affiliation of persons, spaces, communications and inscriptions into a durable form" (2008, 65).[3] Such a perspective understands durability as a result of recurring inscription processes that iterate power stable into a strategic, stabilized node. Inscription is a "way of acting upon the real, a way of devising techniques of inscribing it in such a way as to make the domain in question susceptible to evaluation, calculation and intervention" (2008, 626). Mobile media in many different ways fits into a larger inscription scheme: it acts upon the real, intervenes, gets evaluated and calculated, and at the same time triggers another will to knowledge and power: that of the sovereign state to control and surveil.

To briefly summarize: mobile media channels regimes of practices for the individuals who in their performance produce the reality of those regimes. But this performativity cannot be determined and the effects in reality are manifold and vary, so that mobile media as well serves purposes beyond, or in contradiction to, programs of governmentality. Still, mobile media 'serves' as a governmental programme for the addressability and enhanced productivity of the governed.

What is left for now are questions of method. On the one hand, the relation of empirical research and an analytics of government, on the other the applicability of governmental studies within non-Western societies.

---

3 Actor Network Theory can contribute to an analytics of government, as it also follows the actants, that are both human and non-human. See Conradi and Muhle (2011), Röhle (2011), and Seier (2011) for the potentialities of a Foucauldian ANT and its limits.

Empirical Statements and Governmental Studies

The body of material that usually is applied in this field of studies is comprised of a discourse analysis, which relates to documents that describe and prescribe programs of government. Less common is the application of interview statements within a governmental studies frame. Bröckling et al. argue why:

> The focus is on the interrelations between regimes of self-government and technologies of controlling and shaping the conduct of individuals and collectives, not on what human beings governed by these regimes and technologies actually say and do. (Bröckling et al. 2010, 13)

An actor centrism contrasts with this focus of interrelations of entities and artifacts, discourses and subjects.

For instance, a research interested in biographic assessments, individuality in statements, bottom-up theory approaches (like grounded theory), and the methodology of working as careful as possible towards any interpretation of the interviewees' statements clashes with a framework that draws on discourse, dispositifs, and in general non-personal residua.

It is therefore important to set the parameters right for the status of the interviews. These are *expert* statements, there is no ethnographic or anthropological interest in them. Neither the interviewee's biographies nor the interviewees' own narratives are highlighted or play a significant role. Moreover the interviews focus on the *how*: how mobile media is used in protests and social struggles, how it renders empowerment. How within the paradoxes of liberal governing, mobile media becomes surveillance media.

In a reciprocal move, the interview statements have guided the writing and the unfolding of arguments. This was done in a recurring reading the statements against theory. Therefore, some statements are directly interwoven within the text, others contribute to the larger argument, while again others are introductory statements to sections or single arguments. As they are referenced and discussed as expert statements, they are part of an analyzed discourse on mobile media.

Many interview statements consider local issues. Many do so in the hope to make the practice and interrelations better understood. The telling of contexts and situations refers to the goal of situating a practice within a context of governing and thus to make it translatable into non-local reports[4].

Methodologically, expert interviews can be referenced in the same manner as documents and other parts of specific discourses. Because it is not the person as such that is of interest, but her function in an institution or organization, which can contribute in semi-open interviews to the field of action. Experts have operational knowledge (Meuser and Nagel 2005). Statements that reflect the practice and use of mobile media are not academic in nature. They are single expressions of a discourse, which is prevailing around mobile media.

Still, some issues of empirical research within governmental studies remain. I do not think there is a 'clean' solution to this. A remaining ambiguity of the performance that these statements can achieve within governmental studies goes along with this work. But the accounts here are rarely personal accounts, and they commonly describe practices with mobile media, which again echo discoursive formations and the established horizons of truth. The interviewees are part of groups or organizations. In this regard they contribute, not as single actors. It is not the interviewees' own life that is of interest here, but the effects of mobile media as a political technology on activist's practices, and thus processes of subjectivation resulting thereof. In that sense, the statements can highlight governmental practices, whether as statements of (self-)governing, or as statements that proclaim resistance to certain aspects of governing.

Resistance then is addressed in two senses: once as a kind of appealing resistance, one which locates itself within governmental framings to update governmental practice towards aspirations of the society or specific group, and once as a kind of 'refusing resistance', which does not address the state or corporations, but claim autonomy from the state as its strategy. These statements refuse to address the state and the official designated space of politics *and* society. Both are kinds of resistances that face the illiberal and disciplinary modality of rule (see chapter 7 for a smarter distinction).

The documents provided in chapter 8 about surveillance operations on the 'militant group' are examples of practices, which have rationalized themselves in the logic of endless suspicion. It is the general lack of access to such documents that

4 This is not a contradiction: context serves as a medium of comparison to show differences and similarities alike.

reduce the discussion of mobile surveillance media practice to one case. More would be useful. Many interview statements also provide details, but these details do not originate from the surveillant party. This is a difference in quality.

To summarize: the intermingling of interview statements with governmental theorizing remains ambivalent. It is of a limited resilience, but on the other hand, is seen as uttering of a discourse about subjectivation; about processes that rationalize the political technology mobile media. It is an informed and detail-prone way of representing governmental and disciplinary practices and techniques in a field of contestation. The interviewees' statements, I propose, may be read as statements delivered from an informed, experienced and reflected position. They sum up to an agglomeration that meets an explorative space in which the governmentality of mobile media is expressed. They allow a granular and precise discussion thereof.

Before entering the specificities of mobile media as a political technology in further detail, another matter has to be discussed: to what extent can an analytics of government that relies on the concept of governmentality be applied across the globe?

Governmental Studies, Eurocentrism and 'Postcolonial' Societies

> The price of a cell phone is not an obstacle, maybe the half of Oaxaca's population has a cell phone
>
> Blax / Oaxaca (Mx)

Governmental studies do not refer to a normativity or provide tools to describe ethics. They are "not concerned [...] with making judgments as to whether and why this or that policy succeeded or failed [...] to analyse what one might term 'the will to govern' is not to participate enthusiastically in it" (Miller and Rose 2008, 29). Since they derive in a long line from post-nietzschean genealogy, they do not have a point of reference in meta-historical terms, no teleology, or a narratives of emancipation (Dean 2010, 75-86).

They show *how* rule has rationalized itself as acceptable and how it came to be regarded as such through a huge and heterogeneous arsenal of performative practices, techniques, and technologies. For governmental studies *there are no universalities.*[5]

Governmental studies much more follow the idea of what can be called de-objectivation of objectivated settings. With this I mean the prospect to make comprehensible and thus question how it came to a formation of diverse dispositional settings which appear natural in their everday manifestations. By doing so, the frictions and power-vectors become visible and traceable, which have led to the present condition and its societal setting.

To give an example: how torture is perceived in discourse is an apparent example, as it differs in meaning and function throughout its historical manifestations. The rationalization of torture itself is bound to historical criteria:

> The ceremony of public torture isn't in itself more irrational than imprisonment in a cell; [...] One isn't assessing things in terms of an absolute against which they could be evaluated as constituting more or less perfect forms of rationality but, rather, examining how forms of rationality inscribe themselves in practice or systems of practices, and what role they play within them – because it's true that 'practices' do not exist without a certain regime of rationality. (Foucault 2000b, 229-230)

And if one thinks about the reappearance of torture within dispositifs of securitization in late liberal societies, most prominently after 9/11, the way how it is conceived in general is with disgust and embarrassment. It functions differently today than it did in the "classical age", before the bourgeois revolutions, because today's rationalities – the contemporary production of truth – provide different normative values about the political technology of torture.

Then, if governmental studies ask "how" programs and practices are established (Miller and Rose 2008, 6), and if they do not deal with "subjects then, but subjectivations, as a mode of action on actions" (2008, 8), I do not see an intrinsic reason why they should not be used to analyze non-Western and, more specifically, former colonial societies, or postcolonial societies. What makes the case for this task

5 Or, being asked what liberty might be, Foucault puts it simple: "Liberty is a *practice*." (Foucault 2000d, 354; emphasis in original)

even more convincing is that "it seems impossible to even grasp the formation of the modern European state and its biopolitical dimension without accounting for its colonial dimension" (Saar 2010, 48).[6] The limits are set, as focusing on governmental programs one applies an inherently Western diagnostic view onto societies in a way that prevailing indigenous knowledges are pushed to the back of analysis. One has to be aware of the heuristics involved.

In the context of this study it is a potential source of analytical trouble that governmental studies demand an argument in terms of location and historical setting. The interviews referenced here are statements from more than 10 countries, that are as different as Mexico, Pakistan, and Japan. But on the other hand, the larger subject discussed here is a global affair technically, as well in large parts governmentally. Although mobile media always finds its own expression and specificities of regimes of practice in local circumstances, no universality of mobile media regimes exists as such. Still, as a programme of government it is rolled out to achieve briefly the same effects: *a less constrained flow of things, signs and people*. In many instances, mobile media bridges deficits of other media, which further contributes to the empowerment of the self. With regard to China, Tibet and Burma, Nathan Freitas names what mobile media steps in with.

> It is mainly used to tell these stories often from places where traditional media isn't accessible.
>
> Freitas / NYC

By claiming the applicability of governmentality via mobile media in a global context, the limits already on the technical availability of mobile media are reflected. It goes without saying that mobile media as discussed here, is not rolled out everywhere.

> In Agre we have no mobile phone signal. That's why we choose radio.
>
> Caminaghi / Campinas (Br)

My general claim is that research that focuses on how signs are made to circulate, how the flow of signs is fabricated,[7] and in what way mobile media supports

6 See also "Race and the Education of Desire: Foucault's History of Sexuality and the Colonial Order of Things" by Ann Stoler (1995).

7 It is interesting how Innis arranges his research:"I do not intend to concentrate on microscope studies of small periods or regions in the history of the British Empire [...] Nor shall I confine my

this, and how and when. This research can in part neglect local criteria that deeply reflect local cultural structures, as a straight-forward universalization is not the goal as such. Mobile media can be understood as a vector of powers that leaves traces everywhere. Some are more universal than others.

Analyzing "Development"

Governmental studies trace political reason and rationalities in Western, and by and large liberal societies (Foucault explicitly analyzes post-war West Germany, the U.S. and, not surprisingly, France). An application onto other societal constellations in different regions encounters problems: as it is precisely the refusal to draw on universal concepts that allows this analysis to work, such a study cannot but only look at specific historical and local settings. Bröckling et al. note that in "reconstructing local orders of knowledge and regimes of practice of varying scope, studies of governmentality deconstruct the idea of universal reason" (2010, 11). Studies of governmentality are thus localized studies, they are situated. Nonetheless, I propose that they can be situated in a global regime of practice, which expresses itself with different facets locally, like the global practice of mobile media. It allows the analysis of other "forms of government, including authoritarian rule, and is in line with the way in which Foucault's own understanding of governmentality was developing" (Death 2011a, 15).

It is evident that Foucault draws on material of European history in the development of his concept of governmentality. At the same time though, Dean has a convincing point in carefully broadening this to an analytical grid with a perspective "that seeks to connect questions of government, politics and administration to the space of bodies, lives, selves and persons" (Dean 2010, 20). Here it develops into a study of any form of governing that channels regimes of practice, of how the three forms of rule, the triangle of sovereign, discipline and governmental management, relate to each other. Evidently, possible different conglomerations of these forms of rule manifest themselves differently in different contexts.

> Also Musharraf was really a victim of his own policies because he liberalized the government controlled media. And as a result, one

---

interest to the British Empire as a unique phenomenon" (Innis 2007, 23). By way of understanding communication as in a "crucial position in the organization and administration of government" (2007, 23) Innis suggests to see communication studies invested with a wide explanatory power, which is limited by cultural factors only to small degrees.

> of the things he did, there was a lot greater media freedom than there was in 1999 when he took power. And you can certainly say that he was a dictator, but this doesn't mean that everything he did was bad. But because of that, when he did have his second coup as they say in Pakistan: when General Musharraf overthrew President Musharraf, because he was both President and General at that time, the media would not sit down and take it.
>
> Theriot-Orr / Seattle

In a simplified way, one might say that the litmus test if governmental studies can contribute to the analytics of governing in a specific region is provided by the critical question of the exercise of freedom: "Government is an activity that shapes the field of action and thus, in this sense, attempts to shape freedom. The governed are free in that they are actors, i.e. it is possible for them to act and think in a variety of ways, and sometimes in ways not foreseen by authorities" (Dean 2010, 21). If the governed are not in the position to act in ways they choose, the extent of governmental rule is limited and the dominant model of rule is structured by a different mode, by sovereign rule, which demands obedience.

> Surveillance in the Philippines is different from others, here it is quite primitive. They also use the cell phone, so people in my hometown would text the military that I am home, others would text me "get out of the house", so it works both ways. For a while I thought it is funny to play hide and seek with the local militias, but the next step, after locating your exact whereabouts, they are at you and you just disappear like a bubble in the air.
>
> Reyes / Manila

But even in most repressive modes of rule some 'quantity' of freedom remains which materializes itself by practices such as using YouTube to disseminate proof of brutal assaults.

> Since last year, a few of the activists especially involved with the Teacher's Union [APPO, O.L.] tried to use the cell phone with video recording capabilities especially to document the events of the police repression and to send it immediatly to a webpage, sending it to You Tube
>
> Blax / Oaxaca (Mx)

The vector of technologies of dissemination, the machines of circulation, to my understanding, preserve choices and in this sense demonstrate that liberal regimes of practice, once established, cannot simply be nullified.

Postcolonial Societies

Postcolonial theory has been inspired by a "rethinking of power" and "the concept of discourse and the power/ knowledge nexus have found particular resonance in analyses of colonial and postcolonial relationships" (Abrahamsen 2003, 199). Carl Death has shown the applicability of such an analytical approach towards agency in some African contexts. I agree with his claim that

> Whilst using governmentality synonymously with neo-liberalism does indeed restrict its applicability, the use of governmentality as an analytical framework or grid for understanding power relations more generally [...] renders it more broadly applicable. (Death 2011a, 24)

This perspective, to apply governmentality studies beyond their genealogy from neo-liberalism and thus enriching the analytical tools to understand rationalizations of governing more generally, can be exemplified if one looks at knowledge regimes. Human rights activists in Brazil or India, which fight for the recognition of indigenous people and their needs, fall within the dilemma of liberal rule: whom to include into civil society invested with rights and whom to leave out of the picture. What is more, indigenous struggles for recognition of their own concepts of property or commonality can be understood as a competitive endeavor to liberal rule, which itself is based on Western concepts of property and enterprise. The trajectory of Western rationalizations disqualify native knowledge. "[A] more public circulation of reason would serve to undermine and break down the supports of native knowledges, to disqualify them." (Scott 1995, 209-210) Liberal rationalities competes with non-Western rationalities and regimes, it tries to nullify other notions of knowledge by way of its mechanism of centrifugality. The area of development aid demonstrates many examples of how governmental management is being invoked in new fields and areas. For a better understanding of this argument, an article about the dissent of the Botswana Bushmen, against their relocation, is instructive as an example in showing how political technologies, such as participation and integration into government, are deployed in this region. Odysseos refers to Foucault's framework of government in his analysis:

> In understanding the Bushmen as merely subsisting in their traditional ways of life within CKGR [Central Kalahari Game Reserve, O.L.] and

> insisting that providing them with basic services for health and education therein was non-viable, the GoB [Government of Botswana, O.L.] has sought to control and enhance their activities, to promote more productive paths for them to traverse. To be viable and to transcend beyond mere survival, the Bushmen must be enmeshed in the governmental project of development. (Odysseos 2011, 444)

Control by means of enhancing activities: offer possibilities, nurture with choices, integrate into circuits. Odysseos gets to this point only some pages later:

> And those governmental agendas wish to 'elevate' the Bushmen out of 'mere subsistence', develop them as increasingly self-reliant and sedentary subjects engaged in agro-pastoralism, but also open and welcoming of pastoral interventions in the economy, health, and education. Their production as 'governmentalizable' subjects, then, involves generating new desires and interests in different (settled) ways of life. (Odysseos 2011, 447)

The production of 'governmentalizable' subjects has always been of central concern for governmental management. And a means to do so has been, and still is, "to put in place a public sphere in which only certain kinds of knowledges and not others could circulate with any efficacy" (Scott 1995, 209). Like civil society "in which fluency in these knowledges [...] would be a condition of participation; and in which participation would be the only rational and legal way of exercising influence in what now counted as politics" (1995, 210). In the Botswana case a contemporary state is struggling with the production of its population as governmentalizable. In effect, this demonstrates how broad the application of governmental studies can be.

Colonial Governmentality

Kalpagam writes about colonial India as a case where a new regime of norms was established that replaced, with and *without* means of force, other variants of economy:

> Colonial governmentality, in thus setting its own norms, disrupted the moral economy of peasants that was also governed by norms, albeit

> different ones. Such substitutions were clothed in a discourse of bureaucratic rationality that framed peasant behavior as 'fatalistic' and 'superstitious'. (Kalpagam 2000, 433)

The trajectory of economic inclusion into a global sphere of interchangeable goods, and the establishment and enforcement of the market regime as the place of exchange of societal relations remains an ongoing venture. Such a program from its inception to its enforcement needed a milieu in reality that only governmental subjectivities could provide, even within the populations of colonial rule. Scholars of colonial and postcolonial government describe the production of an economical environment to be most important key understand what made colonial rule successful. "Colonial governmentality in India thus made it possible to conceive the Indian economy in the modern economic categories of income, wealth, production, exchange, distribution, and consumption." (Kalpagam 2000, 421) Alpha from Mexico echoes the connection between technologies and the establishment of capitalism:

> As an anthropologist I have realised that technology is very much linked to capitalism and creates needs. So, this idea of progress linked to technology is not attractive to indigenous groups who are different in their communities. For them, communication means sometimes wandering 30km from one point to another. I relate technology mostly with the question of consumers.
>
> Alpha / Mexico City

The target of "modern power is not so much the body of the sovereign's subject [...] as the conditions in which that body is to live and define its life" (Scott 1995, 199; see also Lemons 2008). This is the milieu of the apparatuses of security. There is a very complex nexus of different kinds of rules present in colonialism, and "one of the things the new game of politics came to depend upon was the construction of a legally instituted space where legally defined subjects could exercise rights, however limited those might have been" (1995, 208). For Scott, who systematically included Foucauldian thought into his postcolonial analysis, a new form of power emerged;

> which was concerned above all with disabling old forms of life by systematically breaking down their conditions, and with constructing in their place new conditions so as to enable – indeed, so as to *oblige*

> – new forms of life to come into being." (Scott 1995, 193; emphasis in original)

These new forms of life were the ones that could be administered and governed *although* colonial rule was first of all sovereign rule, which possessed the bodies – and only in the enforcement of a market regime the subjects gained rights as market participators. Scott thus wants to bring to light "the formation of historically heterogeneous rationalities through which the political sovereignties of colonial rule were constructed" (Scott 1995, 193), and he engages in different power perspectives and regards "colonial modernity" as a "discontinuity in the organization of colonial rule" (1995, 193).

This is the emergence of

> a colonial governmentality – in which power comes to be directed at the destruction and reconstruction of colonial space so as to produce not so much extractive-effects on colonial bodies as governing-effects on colonial conduct. (1995, 204)

The production of conduct aims at bodies *and* subjectivities. Although something of its own kind, one can understand colonial rule a heterogeneous mix of sovereign, disciplinary and governmental management. When "colonial power came to depend precisely upon the systematic attempt to intervene at the level of [...] 'society itself' " (Scott 1995, 207), the modal way of governing resembles European rule in that a political economy became the horizon of truth for both. When economy was remodeled, so that "only principles of economic authority and distinction to be allowed were those defined by the abstract and self-regulating demands of the market", the narrative is very familiar to that of (neo-)liberal ideologies, as it "operated not on such aggregates as caste but on *individuals* responding only to the rational or natural pressure of want and self-interest" (1995, 211; my emphasis). Colonial governmentality established a new rationality, "one which the colonized could accept or resist, but to whose rules they would have to respond" (1995, 212). The response, one may conclude, was part of a governmental programme in which "the native was made to work upon himself; he was now conceived of as a productive agent" (Scott 1995, 213).

This short detour about colonialism provides in particular the argument that governmentalization of subjects in a specific way appeared not only in liberal Europe, but in parallel in the colonialized regions.[8] It even was centered around the production of conduct and self-conduct. It promoted productive agents, which would engage in exchange – just like today the Bushmen of Botswana are addressed and interpellated, although, I hope, to a much lesser degree by force.

Today, the topos is 'globalization' when it comes to the subjugation of spaces under one economic regime. Mobile media is not only a signifier of a global sphere, but as well in practice facilitates regime change towards global patterns and connections:

> If globalization is understood very broadly as an increasing porosity of boundaries and changing experience of time and place [...] mobile phones contribute to such processes by facilitating a multiplicity of relationships in areas that used to be relatively isolated. (Tenhunen 2008, 515-516)

Mobile media acts, and has effects, as an agent of globalization by invoking the individual, making it literally perceptive to calls, which is just a way of interpellation, and facilitating relationships beyond traditional ties. This is what Sigley points out:

> A number of the key administrative states undergoing 'governmentalization' were also concomitantly administering or acquiring colonial possessions. It is, therefore, impossible to do justice to the dimensions of this process by which the 'conduct of conduct' becomes preeminent without considering the global dimensions of its coming into being. The governing of subjects at home and abroad are closely intertwined. (Sigley 2006, 491)

Further, he reminds us that the liberal image itself could only have been constructed through its exterior, as it was "to a large extent reliant upon the characterization of non-liberal, that is, non-Western, societies as liberalism's Other" (2006, 490-491). Thus, outside of the European discourse, this 'Other' has been, as Scott argues, subjectivated in a governmental manner, too, albeit in a different configuration of the programs, and produced with other means.

8 Of course, it would be necessary to show this in more detail and for different countries, not only India.

Katherine Lemons (2008) reminds us about the modifications necessary for an application of an analytics of rule onto postcolonial spaces. She argues that it is possible and fruitful to analyze current Indian rule within a frame of governmentality, but rightly warns that modifications are necessary in regard to religion within this analytical grid. Especially since religious law is codified in contemporary India in the fields of marriages, she argues, governmental studies need to be modified for an application in this context. This is convincing for the study of these societal fields, as it shows the limitations and necessary modifications for specific fields. Such modifications need to be applied on particular fields of study. Mobile media programs, and the subjectivities it produces, can be analyzed without considering religious laws and their effects on subjects.

Nancy Holland (2002) argues that without pastoral power Foucault's framework does not make much sense, thus it can be applied only to regions with a dominant Christian culture. My argument is that there is no direct, necessary connection between these fields. Pastoral power as a functionality is not limited to areas of Christianity. In the 21st century pastoral power has become a power of telecommunication circuits. Pastoral power has, maybe to lesser degrees and maybe in a different way and form, implicated societies that are not dominated by Christian heritage. In a sense *mobile media is a bearer of pastoral power*, a medium of and for pastoral power, that engages users in chatting and jabbering. Pastoral power has modified, just as it has in the genealogy described by Foucault. Pastoral power of the 21st century, in the sense of the history of this present, continues to develop into new strains of social relations, and most prominently and naturally within the realm of individual communication.

The most dominant force of its limitation is of economical nature.

> Problem is that most people don't have credits for the phone.
>
> M / São Paulo

My impression from a poor country like the Philippines sustains this further development of pastoral power. Undoubtedly one should not describe the Philippines as a country of dominantly liberal rule, but still there are aspects that can be seen so. A convincing example is the massive use of mobile media and the Philippine people's affection towards SMS.

The mobile phone in a country like the Philippines introduces far reaching features, as Pertierra states:

> While lacking domestic space to develop private liaisons, many Filipinos have turned to mobile phones as another alternative. Texting has become the major way for most Filipinos to cultivate a network of acquaintances known only to them. [...] Several telecom and other services exist to cater this need. While contacting strangers was possible in the past, texting provides the anonymity, privacy and convenience not hitherto available. (Pertierra 2006, 6)

To check the availability of mobile media, I bought many SIM cards during my stay in the Philippines and tested the many possible ways in how to load them. It could always be done without registration, without a bank account or credit card payment. Such *is* a technique of liberalitiy: to include the biggest possible part of the population in telecommunications. The effect of such a liberal economic regime changes societal relations in general, as new ways to connect weaken traditional modes of social interaction. The individual is addressed as having choices upon its own considerations. The address space of the mobile is a space of choices.

In that sense I doubt that an analysis of mobile media use from a governmental perspective has to restrict itself to European and U.S. terrains. Liberal rule is not once laid down, activated, and then exists, but must be understood as a process of permanent re-evaluation and modifications of regimes of practices. Global mobile phone networks therefore refer to a sedimentation of an aspect of liberal rule. If, and how, this is more than a sediment depends on the terms of use, the roll out, the way it is used locally, how payment can be done, and finally on the many illiberal measures that try to minimize the risk of this production of freedom.

Additionally, a hint from a MST worker in Brazil points to a strength of governmental studies:

> `Laws in Brazil are not important.`
>
> MST Workers / São Paulo

Whatever laws are written down, whatever official 'government' exists, it is the regimes of practices that establishes the realities of governing. The existence of democratic institutions, the existence of a constitution or of a civic code are never enough for an analytics of rule, because of interest is *how* the process of rationalization succeeds. Mobile media is one "how" in a long chain amongst many other chains.

Mobile Media in the Global South: Hook Them Up!

Mobile media's connectivity reach is far beyond that of the internet. The internet failed to include large parts of populations into the circuits of addressability. To describe the internet as a facilitator of governmentality on a global scale is dangerous. Even institutions are not yet well connected in a country like Mexico:

> Here in INA [National Institute of Anthropology and History, O.L.] we have for 300 computers a 4 mbit connection.
>
> Alpha / Mexico City

Also, mobile media connectivity has limits. But these are far beyond the limits of the internet. Still, some areas will maybe never be connected to the global infrastructure of GSM. So satellite connections help to communicate at a basic level:

> If you take something like the satellite uplink for the internet. This has been incredibly important. Apart from being able to give the community access to the internet and then have to maintain contact if you are in some remote area in Guatamala, where it takes three days to go to and there are no other communication means. And maybe they would have one computer maintained to get onto the internet, then you can talk to them and help them fix stuff remotely without having to go there. Also get feedback by email from that community instead getting feedback only when someone goes out. That stuff is hugely important.
>
> Anonymous1 / Mexico City

But this is not a mass phenomena. In effect, its exclusivity limits new regimes of practices and subjectivities to be teletransported.

For specific practices the problem of connectivity can be bridged with analog means.

> There was a protest march from India back to Tibet where a number of 12 monks and other lay people were marching back to Tibet. And so they had these little cameras and they would document every day and they get into their camp each night and use the satellite to upload the footage. Or they would put it on motorcycle and the guy would ride to the nearest internet cafe.
>
> Freitas / NYC

Looking at the strategies and efforts of mobile phone companies and their partners in the Global South or 'developing world', it turns out that the mobile phone is a key facilitator of developmental management. My argument is that mobile media in many instances is part of a 'development' plan which facilitates governmental practices. At the contemporary frontier of governmentalization mobile media is rolled out where former means failed due to different reasons, and often because the state has been too weak to penetrate the population with other means of communication. Mobile media works as an economic productivity factor on micro levels. Jonathan Donner, a mobile phone researcher for Microsoft, states:

> M-banking and m-payments systems, such as M-PESA in Kenya and WIZZIT in South Africa, are among the most exciting applications: they are surprisingly simple, often using SMS or other basic interfaces, and they have demonstrated broad appeal, even among traditionally "unbanked" populations. In Kenya, at least four million accounts have been registered by M-PESA in its first two years of operation, a number that is rapidly approaching equivalence to the number of formal bank accounts in the country. (Donner 2009, 96-97)

Mobile phones integrate many more people into the liberalized financial system and make them partakers of the market. Of course, changes in the banking model occur in parallel, as the companies that provide these services are not banks. As the costs for accounting and administration are lower, smaller balances already produce a good return of investment. Mobile phones are able to integrate larger parts of the population whose productivity has not been brought under the rule of a regulated economy. The incentives had been too low so far. Mobile media changes this scale.

It is striking to see very similar procedures at work today as in 19th century colonial India "in the myriad of economic transactions that enveloped the colonial economy" (Kalpagam 2000, 421). This time, the addressability the phones provide facilitates the circulation of money.

Mobile media becomes part of rendering instrumental rationalities and as such it is a technology just like "all those devices, tools, techniques, rendering, personnel, materials and apparatuses that enabled authorities to imagine and act upon the conduct of persons individually and collectively, and in locales that were often distant" (Miller and Rose 2008, 16). The distance between the subjects, and between the subjects and authorities, can be bridged swiftly and cheaply with mobile phones. Mobile media governs literally from a distance, while it is carried at the body.

> Many people in Brazil have a cell phone but don't have a landline, because there is a fixed cost for a landline. And you can get a cellphone for some 40 to 50 real with the SIM card. That's cheap enough for everyone. I have friends in the homeless movements, street people, who have mobiles, too
>
> M / São Paulo

Street people, I would argue, are the natural targets of mobile media, once the costs are rationalized. In a sense, street people, without proper residency, are the natural beneficiaries of mobile media, as it comes with an address of its own for people without an address.

Donner characterizes mobile phones as market enforcers. "[M]obiles improve the efficiency of markets, enforce the law of one price, reduce waste, and increase productivity" (Donner 2009, 92). Others disagree as "it is far from established that the mobile is yet affecting income distribution in either direction" (Coyle (2005, 5); see also Rao and Dasgupta (2010) and in regard to Chinese low-income population Wallis (2011)). Donner goes further, as he argues that the societal relations a population establishes and nurtures within the realm of mobile media, "later can be used for transactional and 'productive' activities of interest to the economic development community" (Donner 2009, 97).

How to return profits under different and harsher conditions are the key elements in this roll out of governmental management technologies. One example is "the use of a pre-paid chip or phone card, which can be used with a handset belonging to someone else. This arrangement allows the user to place a call using someone else's hardware and battery power, but with their respective pre-paid mobile phone chip or card" (Sinha 2005, 10).

The means chosen to hook up poorer populations to mobile telephony are impressive: corporations such as Tata in India suspended "charges on mobile calls within the company's network for a two-year period" (Tenhunen 2008, 519) to promote mobile telephony amongst those that otherwise could not afford it.

> The maximum people that we are connected with are not on the internet, many people are not even literate. But they have a phone.

> The main demographic that we are relating to, like the fisher people, all these people are not on the internet, maybe one person, some leaders. But all the others are not. But they all have phones. Everyone.
>
> ManasaSarovara / Bangalore

Mobile phones even have become *direct* means of payments. The ability to call has literally become a currency as "pre-paid phone cards and chips are becoming a new form of currency in some areas – where communication capacity is a commodity that is in high demand" (Sinha 2005, 11).

Communication capacity can be traded, but there are interesting loopholes within this economic regime, such as "beeping" instead of calling someone to save costs (Tomitsch, Sturm, Konzett, Bolin, Wagner, and Grechenig 2010, 3; Sinha 2005, 12-13). This practice is widespread on a large scale in so many countries that it poses problems for the telecommunication companies. In India, nearly 40% of the mobile phone bandwidth is used for 'missed calls' (Rao and Dasgupta 2010). In Bangladesh, in 2008 "mobile penetration is roughly 23%, what sets Bangladesh apart from most developing nations is the extremely high network coverage of 98%." The result is an increase in a free use of the infrastructure by letting the phone ring only (Geirbo and Helmersen 2008, 77). The free-of-charge method fits illiteracy as "[f]inding a number in your contact list, hitting the 'call'-button and then, once you hear the call signal, pushing the 'end call'-button is far easier than composing an SMS, especially if you are illiterate or not familiar with the alphabet on your phone's keypad" (Geirbo and Helmersen 2008, 78). As well practical and entertaining, romantic motives are present in the phenomena of 'missed calls' (Geirbo and Helmersen 2008, 78).

For many parts of India, mobile telephony exists where electricity does not.

> Mobile phones are very useful in a country like India, because many of these movements work in places where electricity is not there and no access to internet. The emails are only for the urban situation, but for the rural it is the mobile where you can communicate on the phone using voice but also SMS, which became useful in the way that if anything happens anywhere in the country you create a text message and send it across the country to all supporters in the villages and the cities, and it is easy and quick to organise a support
>
> Kumar / Delhi

And he continues more generally:

> Something like 40% in India have mobile phones, also deep in the rural areas. The penetration is on a high level, except in the tribal villages, where people don't use technology so much. But many of the people in movements are too poor to have phones. But the main contact people always have a phone
>
> Kumar / Delhi

Vijahan in general shares the assumption of this huge dissemination of mobile media, but he also points to the problem of exclusion for those that are not connected.

> Sometimes when you use texting and email you forget that there is a large part of the population that is outside of this network. And for a country like India the masses of this country are beyond the reach of technologies now. Many times, typical middle class people state "even the fishermen have mobiles now". This is a class problem for them, but at the same time it is only a very privileged few among the working classes who have access to this. Even mobile. It has a great range in rural areas. There are hardly any places left where there is no network coverage at all. There are some left. But this does not mean everyone has a mobile. If you go to indigenous people, they don't have it. In whole villages not even a single person has a mobile phone. The land gentry, who has land, socially middle class, there every household has a mobile. So this is a class/caste difference that exists in terms of technology reach. For people who have traditionally not been able to get hold of any technologies, even a second hand mobile, they don't pay for making phone calls. It is just a status symbol. The middle class have huge access to it. To realize that just by communicating on technological basis is enough for your dissemination is wrong and can be an extremely dangerous approach.
>
> Vijayan / Delhi

To conclude: I have shown that efforts have been huge to further disseminate mobile media globally, with high success as the penetration rate is extremly high. Mobile media facilitates forms of exchange, the flow of signs, even in the rather limited vocabulary of the 'missed calls' amongst poor populations. It inscribes itself into a governmental mode of rule that encourages the subject to express itself, to enter into communication, and offers the disruption of traditional social bonds, such as segregations. It allows communication free from ties and as such empowers the individual. It reinvents banking (modified to specific capacities). This has effects on processes of subjectivation: the subjects produced adapt to self-technologies as self-management. Mobile communication as such has become a currency: the currency for participation. In both senses: as it integrates into the circuits of addressable spaces, and as it integrates in modalities of payment from a distance.

But there are limits, in many ways specific groups remain excluded from its use.

> People who do not know English face problems with the mobile phone. How to press "on", how to send a message. That problem is still there.
>
> Akkai / Bangalore

One might say that these limits are the limits of governability from a distance by technology. Nonetheless, global mobile phone dissemination is still growing.

The next chapter looks at the ways how resistance and counter-conducts can be seen under a governmental lens and I try to connect mobile media subjectivation to the focal point of protest.

3 Mobile Protest Media: Resisting Government – Governing Resistance

> In 2003 half a million people got to protest against the celebrations of the reunifications. There also was an Asian economic crisis, and SAS, a lot of people died. The atmosphere was intense. The government passed a law then, a national security law, that law includes one article that the government can sue individuals that speak up if they think it is a threat to national security. For example if you advocate the idea of Hong Kong independence, or Tibetean independence or you criticize policies that are sensitive. The government tried to pass that law in that tension. You need to know that before 1997 [the year China resumed sovereignty, O.L.] the biggest debate was about free speech in HK. So that is the background. In 2003 lots of people joined the protests and after that, people got together to discuss how to maintain the power and the people, because of this kind of incident, people got very dissatisfied with the government, so one or two of the government people had to resign.
>
> Freddie / Hong Kong

This chapter shall serve as a brief discussion of a possible framing of mobile media protest. The intention is not to rewrite theories of protests; this would be a different work. But as mobile protest media is a topic here, I argue that a discussion of protests and resistances is insightful as mobile media relates differently to different kinds of protests, and restructures according to their specificities. The amalgamation of protest and mobile media, its semiotic-material ensemble, changes the 'problem' (in the Foucauldian terminology) of protest.

Governmental management is powerful because it enrolls individuals into programs of government without interfering with their formal autonomy. Even more so if this happens, to some extent, without individuals being aware how much the rationalities they themselves orientate at are technically and instrumentally brought into being, to channel the individual's very own practice and choices. Then resistance and protests against 'the government' appear in a new light. The field of protest now grows in complexity since it produces frictions withing governing if power and governing are understood as immanent to the many resistant practices.

> A Tibetan man, who grew up in exile, traveled back into Tibet under saying he was not Tibetan, but he was from India, and did a protest on Mount Everest that was timed with the big event about the Olympics. This was like a year and a half before the Olympics. So, he did a protest, we used a small satellite broadcasted Bgan [a satellite transceiver, O.L.] with a high def video camera and streamed that back to New York. All for just a few thousand dollars. So we produced this video clip which then we provided in NTSC and PAL format and MP4 to the BBC and AP and they all used it. They showed like "Today in Beijing a ceremony" and baba "and on the Mount Everest" and there was this great shot of the people on the Mount Everest and the banner, and so that received huge coverage and then they were detained for four days and we almost wondered where they were.
>
> Freitas / NYC

I have already addressed the different modalities of power. When it comes to protests and resistances, this modulation of power at first seems confusing: how is resistance itself power driven and ordered by power vectors? I argue, that the different modalities of power are interwoven in resistances, but produce different effects. Effects which themselves translate into power relations.

> There is a constant repression, social movements have been criminalized, MST constantly. We suffer a lot of surveillance and attacks. When we had a round of actions, in these weeks of actions, we suffer a lot of attacks. The servers at the school suffered a hundred thousand attacks. We were monitoring this, no breaking in.
>
> MST Workers / São Paulo

Sovereignty intervenes into protest with the limit-figure of the police. This repression, as an effect of power, itself affects subjectivities of participants of protests, in that the self-conduct turns to other directions. Although the 'protest self' is well configured along lines of conduct and discipline, these are expressed differently and have other effects. As I will try to show, the conduct of conduct, as a soft power of rule, is the first domain of powers affected by the experiences of protests and resistances. Mobile protest media, I argue, is a transit passage of such an affection and its effects, that reshapes and reorganizes power relations.

But first, before entering the matter of mobile protest media, the difference of this approach to social movements studies and ethnographic approaches is worth consideration.

Social Movements Studies and Ethnography

There is a large field of literature and research about social movements and protest available, and it is growing. The field of social movements studies is an established field of its own,to which many social science scholars have contributed. From an European and American perspective, Hellmann (2002), Haunss (2005), Herkenrath (2009), and Cox and Nilsen (2007) give an overview on social movements' research. Following Herkenrath, five strains of this research can be differentiated: Collective Behavior Research, Interaction Research (Chicago school), Collective Action, Political Process, Framing (how to reach people), and New Social Movements.

Donnatella della Porta has been an influential researcher in the field as well, particularly on the 'anti-globalization movements'. Amongst her numerous publications, the introduction to social movements with Mario Diani (2006) and the collected volume "Social Movements in a Globalizing World" (co-edited with Hanspeter Kriesi and Dieter Rucht 2009) have become a common source of reference in the field.

Amongst many other works available, Manuel Castell's "The Power of Identity" (2004), which looks specifically at identity construction in a networked society, for example within 'the feminist movement', has become a classic reference.

The approaches of all these works differ significantly with an analytics of governing: none of these scholars looks at protests and resistances with a Foucauldian approach, one which tries to shed light on the political rationalities that organize power. The general narrative within social movements research reformulates aspects of an universalistic image of the state. This image is confronted with a static, essentialist conception of the subject as an entity that struggles for justice. The black boxing of protest subjectivities, as an antagonist to power, thus cannot contribute much to an analysis which foremost understands power relations as ubiquitous, and from there differentiates strains of effects, agency, and affects.

Quite often, modern concepts from the political sciences are transferred to the analysis of social movements. This leads to a reproduction of state-centrism and its concepts within the analysis of protest. By echoing the divide of political and non-political spheres which is established by modern governing itself, the episteme of social movement studies are themselves structured by modern political thought and can only reproduce this matrix.

My claim is that, if I am right, one misses important immanent aspects of protests that are in relation to rule, and immanent with rule, once the separation of the spheres of civil society and the state are reproduced.

Rather than seeing protests and social movements as antagonists to the state (or 'global capital'), and thus reproducing an ahistorical naturalization of civil society, I suggest a different perspective: first, to see protests as a transgressive vector. What is immanent to protests are deterritorializing forces that cannot be accounted for and analyzed with a vocabulary and concepts that hold up a separation of societal fields, which itself is an effect of governmental management. Protests in this transgressive account appear as singularities, which in its course can set aside history[1] and open up an unidentifiable, temporary space of possibilities, or an temporary autonomous zone, as Hakim Bey (1991) called it. Power relations here have organized differently and the overall political rationality of neo-liberal government was overwritten by counter-rationalities and counter-conducts (I will return to these two concepts).

And second, it is crucial to understand protests explicitly as part of civil society, and as such as an element of a milieu which is organized by apparatuses of security in the first place. Civil society, as a transit reality of governing, is the natural realm of resistances and protests. To my understanding, such a historical-genealogical account is not present in current social movements studies.

Another approach to protests is that of ethnography. A recent ethnographic thesis on the 'anti-globalization movements' was published by Jeffrey Juris (2008). Juris acted as a participant observer mostly in Barcelona and participated in demonstrations. He gives an account about the recent horizontal anti-globalization movement(s) that started in Seattle in 1999, had a peak in Genoa 2002, and onwards to the World Social Forums 2005 in Porto Alegre, and to other places. His work is remarkable in the sense that it is a critical insider account with detailed information

1 Or, in the sense of Badiou (2009) *make* history.

of gatherings, discussions and lines of conflict from within mostly anti-hierarchical political networks, such as the People's Global Action[2] or the Dissent! network.[3]

This approach, again, is very different from social movement studies. I see one particular problem within this approach; a problem that is more or less ubiquitous in ethnographic research, and not new either: to whom does it report? Who is interested in a detailed account of how social movements operate, how they organize meetings, how they discuss the use of force? The ethnographic work of protests enriches the knowledge, I would say, not for and not of the social movements themselves. It rather extracts their realities and reproduces a dense description for a different audience. To give one rather random example: "The pinelli social center served as the base of operations for the Black Bloc in Genoa, while militant internationalists slept at nearby Sciorbia Stadium. The first evening planning took place on July 16, where activists decided to separate from the GSF because of the restriction on sticks, rocks, and firearms" (Juris 2008, 182). The ethnographic discipline's imperative of dense descriptions and embodied experience (militant observation), produces great insights that, when it is about protest, no undercover police could do better. I see ethical problems in such a dense description.[4]

Power and Resistance

Returning to the approach here, government is understood as constantly being confronted with resistances, it "is always a precarious affair: it must always take account of the unforeseen and crises of governability" (Bröckling et al. 2010, 19-20).

Thomas Lemke takes a decisive point of view, as struggles and fights, "rather than 'distorting' the 'original' program, they are actually always already part of the programs themselves, actively contributing to 'compromises', 'fissures' and 'incoherences' inside them" (2002, 9). Lemke argues that government cannot be escaped, but that resistances are already always part of government programs. But the important issue is: government can be changed, struggles can be successful. For example, because government needs to adapt, needs to integrate frictions. This is the pivotal point in this concept of government: while on the one side, it comprises many different fields that stretch over the whole of the society, so that the

2 See http://www.agp.org.

3 See http://2007.dissent.org.uk.

4 See a reflection in appendix A in this regard.

impression might occur that resistance has no place in this concept. On the other hand, these distant points are not without power. The net of powers is not top-down. Social struggles, resistances have their say and their effects on government. This is not a one directional model. Repercussions of struggles can restructure large parts of society and even reshape institutions and other sediments of power.[5] But this never means the abolishment of power per se. There are no machines of liberation.

While Foucault did not produce 'the' key text about resistance, he has argued in many ways that power and resistance are mutually inclusive. Power only functions where there is resistance, and resistance only functions in fields of power. Otherwise one could speak not of power, but pure domination, and not of resistance, but pure obedience. "You have to use power relations to refer to the situation where you're not doing what you want. So resistance comes first, and resistance remains superior to the forces of the process; *power relations are obliged to change with the resistance*." (Foucault 2000c, 167; my emphasis) Because power is dissipative and organizes even the smallest elements that comprise human's relations to one another, and even more so, power relations are a necessary condition of societal relation, it is possible to change power relations, so that resistances either change government, or reshape to confront anew. Additionally, in a predominantly governmental society, power of rule is ubiquitous because of the very nature of this style of governmental rule. This is not always wrong.

> During everyday life you feel you can access the social movement in a very quick manner. For example if you know a squat is gonna be evicted you can mobilize people to come within 30 minutes thanks to mobiles. 15 years ago it would take hour. So it is more instantanous for social movements to react.
>
> Anonymous2 / Madrid

In very general terms, the Foucauldian power concept, to my understanding, is sensitive to societal change, as it analyzes the ubiquity of power on very different levels. Thus, it makes power relations quite accessible, whereas classical concepts of power, which rely on a differentiation of power holders and powerless actors, actually cement power lines in their analysis, because access to power relations is then only provided by those who hold power.

5 For a discussion of Foucauldian thought and post-anarchism, see Todd May (1994).

From Resistance to Counter-Rationalities

The first step on the way to conceptualize mobile protest media is 'empowerment' (which I have so far not discussed). Mitchell Dean writes: "empowerment is [...] the normative correlate of the explanatory focus on agency. If human beings are, at least potentially, agents, then they need to be empowered to become so" (Dean 2010, 83). Agency is a potentiality, that unfolds through empowerment. As such, it seems, agency is the most potent and potential resource of government, because it is here that the conduct of conducts connects. Agency is the resort that (neo-)liberal government effectively targets in its endeavor to activate the governed, to conduct their self-conduct. Dean continues that "[p]rogrammes of empowerment are particularly clear examples of those contemporary liberal rationalities of government that endeavor to operationalize the self-governing capacities of the governed in the pursuit of governmental objectives" (2010, 83).

On the other hand, if agency must be activated in the pursuit of self-governing capacities, it is not determined to which objective this empowerment is pursued. Empowerment of agency can have many outcomes, and different, unforeseen practices may be the effect. It is not dependent on *governmental* programs of empowerment.

I suggest to understand protests and collective actions virtually as independent empowerment processes, independent from government programs. They can produce counter-rationalities, independent in the sense that they are, by and large, not envisioned by government programs, although they might even be triggered by them. And at the same moment, they can reshape and reorganize programs of government. But as a process as such, they open the field of possibilities and pass a plane unforeseeable by government programs. This plane is not, and I risk repeating myself here, free from power relations, and also not from discipline, but the "disciplining should not necessarily be regarded as pacification or domestication, but rather as evidence of the co-constitution of relations of power and resistance" (Death 2011b, 432).

Mobile media both criss-crosses this development, as it is a political technology of governmental rule, but at the same time it empowers counter-rationalities and counter-conducts as it, for example, facilitates coordinated collective distributed action (see chapter 5 in this matter).

David Couzens Hoy speculates about the necessary conditions for a critical resistance. His starting point is that "governmentalization is what 'subjugates' or 'subjectifies' people by defining for them the legitimate answers to questions about what counts as a person, what counts as a proper relation to one's gender, or what rights a citizen has" (Hoy 2004, 88). He understands governmentalization as legitimization of certain mindsets over delegitimization of others, which I called the 'channeling of practice'. Critique then is the "desubjugation or desubjectivation of the subject." Thus, it is not about an alternative identity, but it functions "by dissolving your sense of who you are and disrupting your sense of what the right thing to do is" (Hoy 2004, 88-89).

Dissolving one's sense, to follow Hoy's term, is to challenge the political rationalities that guide one. A desubjectivation interferes with such rationalities. This is close to what Barry, Osborne and Rose explain to be the goal of the study of the history of the present, as it would be "a shattering of conventional thought that strikes at the heart of our most taken-for-granted motivations [...] What occurs is a potentially productive uncoupling of experience from its conditions" (Barry et al. 1996, 6).

Counter-X

Counter-rationalities and counter-conducts rely on empowerment of agency. To unleash or deploy processes of de-subjectivation relies on a reconfiguration of subjectivations. Hence vectors of power change – protests and resistances allow for the experience of de-subjectivation; temporal alternatives that, depending on intensity and duration, entail the production of stable counter-rationalities and counter-conducts. These are conducts that escape the political economic conducts of the subject, overwrite them to different degrees and thus allow experience of collective action against the proposed mode of conduct. By this, power relations are restructured, which manifests in the necessity of governing to reform and to redeploy new programs of government. As Foucault stated: governing has to be in concord with desires, as they are too strong to be ignored by government. Counter-rationalities and counter-conducts are sites of desire production which challenge the predominantly economical conduct of governing.

> One example: during the Tokyo City Mayor election, right now we have a right wing mayor. I sent out a message: "add some nose hair

> on the election ads", so a lot of people did that and it was impossible to say who did it all. Noone knew who had sent it, because it is relayed through a lot of people. My personal name does not appear on the adress. I use this mobile mailing list with 500 people. It is all free to send, we have our own server for this shop, which we use for the mailinglist. Another example: when I go to a city I have never been to, then I have a lot flyers that I leave on bicycles with my phone number and "eat the rich" on it. So people call and meet at the station and we start drinking beer. Like this I was going around to every stop all around Tokyo and made a lot of friends.
>
> Matsumoto / Tokyo

If "movements for freedom and emancipation are not located outside or beyond power relations, but themselves produce regimes of subjectification" (Bröckling et al. 2010, 14), an analytics of governing needs to investigate these very own regimes of subjectivation and what their effects are, as they "inaugurate counter-truths centered on the question of how liberated, emancipated subjects are to understand and to shape themselves and others" (Bröckling et al. 2010, 14).

Hence

> an analytics of protest is specifically designed to show how protest and government are mutually constitutive, and thus how forms of resistance have the potential to reinforce and bolster, *as well and at the same time as*, undermining and challenging dominant forms of global governance. (Death 2010, 236; emphasis in original)

Protest is comprised of this permanent duality: to reinforce and to challenge at the same moment, as it is immanent to government, not exterior.

> The politics of the movements is something the state is increasingly worried about. There have been successful actions [...] in terms of stopping of projects, of not allowing land acquisition, of not accepting displacement and rehabilitation. So there is that kind of pressure on the state.
>
> Vijayan / Delhi

Montgomery argues that "discourses of civil disobedience produce norms and imperatives governing the conduct of protesters themselves. The subjects of protest are produced by this discourse, and these subjects often rely on it for their actions to be intelligible" (2010, 102-103). Rationalized by this discourse, counter-rationalities make counter-conducts intelligible.

From Counter-Rationalities to Mobile Protest Media

Taking conduct to the center discussion, the question arises how counter-conducts come into existence. Death proposes to understand counter-conducts as a way

> to develop an *analytics of protest* for the study of contentious politics. [...] it approaches protests and contentious politics not from an actor-centric perspective, but rather orientates itself toward specific practices and rationalities of protest, which themselves work to constitute particular identities and subjectivities through the performance of dissent. (Death 2010, 236; emphasis in original)

Counter-conducts are thus understood as productive as they constitute subjectivities. Their mode of operation is the performance of dissent. By way of performance they produce counter-subjectivities.

My claim is that this is the point where mobile media inscribes itself. As a political technology, mobile media has an effect on the performance of dissent and thus the production of counter-subjectivities. If counter-protests under the lens of an analytics of protest are seen "as assemblages of mentalities and practices, which come to constitute dissenting subjectivities" (Death 2011b, 429), then the practices that produce these dissenting subjectivities are made up in parts by the political technology of mobile media.

As the discourse reproduced here, by way of expert statements, proposes to adhere mobile media as a dominant factor in protests, it follows that the production of dissenting subjectivities is massively influenced by mobile media.

> Months ago, we had a graffiti action. It was on the metro tree cutting here. On July 12th we had a Non Violent Direct Action. We all went with cans of paint and gonna graffiti. The government was going to cut some hundred trees in a park, so we chose those walls, long walls. It was youth and children. We painted one whole section of the wall, in English and Kannada, against the tree cutting. We sent SMS to many people to come. SMS was the primary tool again. The SMS are being forwarded, we sent it to a community radio, they used it. We were 20 to 30 people, which is a good number to do this kind of action. To run away or to protect someone. It worked out very well. In India there is a lot behavioral control, people are expected to dress like this and talk like this. What kind of action you do. People become control freaks. There is a lot of control freaking in this non-profit social movement sector. We are not into

> this style, we don't tell people how to march. A bunch of people just went out, painted the wall, painted what they wanted, we had a really got time, which is very important for us, because life is short. Otherwise the oppressed are oppressing again. We always had the arts involved, which allows a lot of expressions. Even sports, soccer, whatever. We found that in these ways rather than intellectual ways, more in full body involvements, it is better to express.
>
> ManasaSarovara / Bangalore

As mobile media is an integral part in the performance of dissent, its capacity to organize, to mobilize, and to coordinate alters this performance. My general point here is: the facilitation of circulation as such is a neoliberal government program. It is a core trajectory of liberality. The seamless integration of this factor of productivity in the performance of dissent is a sign for mostly two aspects: first, the circulation and flow of signs as such are not a monopoly of one political rationality. Liberal rule made it a trajectory of liberal rule, but this is not embodied in this functionality as such. It can be transposed to suit different trajectories, and apparently, this is the case with mobile media, a concrete manifestation of this (neo-)liberal trajectory.

Its emergence in the full scope of a semiotic-material agglomeration, as a political technology of circulation has no essential core but this circulation. If counter-rationalities can render mobile media instrumental for their counter-truths, no continuity exists between the material and the semiotic, at least at first glance. The performance of dissent incorporates mobile media to make it mobile protest media.

But it is not that simple. So, second, mobile media has its say and do in the course of becoming mobile protest media. While it empowers the dissenting agencies, it so empowers to *specific practice*. It channels practice and favors some over others.

> The common use is sending messages that say "today is a big protest and we can stay there and there", but not to communicate in a strategical way. You can send news in a text message, but it is not conceptualised here in Oaxaca. Maybe in the next years, because also here the use of cell phones is growing. It is a new technology that people only used recently. It is used for individual communication, not for the mass communication
>
> Blax / Oaxaca (Mx)

At first, mobile protest media favors the individual over the group. The individual is its natural anchor. Not only because it is carried around by individuals as a device. Mobile media has become an intimate media, it has become a steady accompanying technology throughout the day and night. This immersion within the intimate radius of the body reshapes practice.

> Whereas telecommunications until then had been bound to fixed places, the mobile phone moves the psycho-social home to the mobile proximity of the body. Thus paradoxically it is the growing independence of social communication from a stationary infrastructure that up values the here and now of the body. (Winkler and Tischleder 2001, p. 102; translation O.L.)

And it is not a collective body that is being up valued, but that of the individual. Likewise, much earlier in its roll-out, it addresses the individual: as a technology of the self, as facilitator for the homo economicus, who can perform her market participation in even greater independence from place and time. It is not a media for the family, like landline telephony used to be often. As I wrote: it integrates individuals into a space of individual, personal addressability.

This can be used as a tactic, if it is rationalized.

> Sometime it is also the tactic that, let's say people have been arrested and detained in a police station, so by any chance you do manage to get the mobile phone number of that station house officer and then you pass it on to all the supporters and everyone starts calling that person. That creates an extra pressure, having people calling from Delhi, from Bombay, it does work. Because also the mobile gives you a thing of person to person communication. When calling a landline number they can say the person is not around. But with mobile the chances are better. This is useful mostly in the firefighting situation
>
> Kumar / Delhi

But this space of addressability works in any direction, and here mobile protest media carries a specific trajectory of empowerment: it divides collectivity. It excludes everyone except the addressed. For 'social media', this specificity is addressed by Fenton and Barassi (2011).

> [P]olitical participation is frequently defined by and takes place in relation to and in coordination with others. Foregrounding creative

> autonomy and the power of individuation may well be appropriate analytical tools for social media, but to do so negates the collective dimension of political participation and thereby dissipates the political properties of the participatory communicative act itself. (2011, 183)

They continue that "in an era of blogs, individual websites, and social networking sites, individual messages are often given the same importance as the messages that have arisen out of the tensions and negotiations of a collective of people" (2011, 187). Much the same is the case with mobile media (and its integration with 'social media' might even spur this effect). The dissenting practice of counter-rationalities face a *technology of individualization*, a technology that divides the social plane into its smallest possible elements.

> Now, if a decision is to be taken, you don't need to sit. The village leader takes a decision by calling four or five people. But there is still this kind of interaction which happens on the phone in contrast to the kind of interaction when you are sitting in person with 10 people. Which is much more useful for the political consciousness to build a team. So the coming of the mobile has an impact on this process. Much of your work is done on the phone. Sometimes you can bypass people, sometimes the information is not passed on
>
> Kumar / Delhi

This is the most prominent site of frictions and where resistance against the very technology itself develops. How can a technology of individualization be changed into a technology of collective action?

> The system that is installed in the squat here is: you push a button and 500 people get a SMS that the squat is evicted. This is for emergency. This is very cheap and effective. By mistake it was activated once, so it works.
>
> Anonymous2 / Madrid

To come to grips with a situated account of protest and mobile media, the next chapter investigates the wave of protests that happened in Pakistan in 2007 and 2008. By providing this section, mobile media's potential of a specific circulation of signs, a circulation from individual to individual, is described. But as well the social means to disrupt this divide of the social plane, for example by intense resending of messages to the complete address book in one's phone, I try to frame mobile media within these specific protests as mobile protest media – where the protest is paired with the political technology mobile media.

4 Mobile Protest Media Case Study: The Lawyers' Movement

In 2007 and 2008 Pakistan was shaken by a movement called the "Lawyers' Movement" or "Anti-Emergency Movement". What started as a protest by lawyers and judges against the sacking of the Chief Justice of Pakistan, Iftikhar Chaudhry, gained momentum after a proclamation of the state of emergency, and finally led to a mobilization of large parts of Pakistan's population. Although the main groups were rooted in middle and upper class milieus, to defend the abstract good of the independent judiciary, the working class and poor people also took the streets.

> In a country like Pakistan where 60% of the people are living below the poverty line and they don't know where their next meal is coming from, for a country like Pakistan to sustain a movement on an ideological base, because you can not see the rule of law, the man on the street thinks about bread on the table, it was an amazing movement.
>
> Zaman / Lahore

Based on interviews, online material, and the available literature, what follows is an analysis that takes a deeper look into the movement, its actors and the context. Therefore, I try to provide a detailed description of a protest movement and its use of mobile media, online media, and the role played by broadcast media. The reason for this case study is the movement's intense use of mobile media. By doing so, this chapter in part is a narration of the events. To provide this, I decided to quote extensively from interviews and articles.

The protesters interviewed about the lawyers' movement and, likewise, the movement itself, differ from all other interviewees in one point: whereas all the others in their efforts aim at social justice, including economic equality as a common goal, and thus understand themselves as protagonists that are antagonistic to the capitalist mode of societal relations, the Pakistanis I interviewed can be considered less radical in these terms. The fight for an independent judiciary is not in itself targeting capital means of societal relations. Some commentators even argue that the lawyers' movement was mainly after a status quo for a milieu of professionals.

> The lawyers' movement in Pakistan, for example, while employing the devices of people-power movements and using the language of democracy, was not primarily about shifting power to a wider spectrum of the population. Rather, it was about resisting the reduction of power of a proportionally small elite that the military regime had initiated. This was not an attempt to bring about a great new South Asian democracy – although a strong and independent judiciary is associated with stable democratic regimes in Western Europe and North America, so it was not incompatible with a potentially more far-reaching and broad-based pro-democracy people-power movement. (Lyon 2010)

But the specificity of this profession, being the ones that either act as representatives of the law as judges or as lawyers, makes them agents of an abstract goal as well. In my encounters with lawyers and students I was often confronted with a very strong nationalist wording, most prominently towards India, and an underrepresentation of poverty in the discussions that took place. Nonetheless, to present the lawyer's movement in more detail, although it clearly is not representative of the general spectrum of activists I interviewed, can be argued for:

First, in a more general sense, mobile media is a tool for empowerment and protest agency, irrespective of the goals it is used for.

Second, for an analytics of rule, as outlined above, this movement in Pakistan is a convincing case, as both sovereign and governmental modes of rule can be separated and it can be shown how these modes of rule can be in contradiction and thus have the potential to clash. Especially, the invocation of the state of emergency as an act of sovereignty allowing the state to shoot protesters in its boldest manifestation (even though this never happened overtly), is in direct contradiction to both the massive use of mobile media as an facilitator of liberal rule, and the protests for an independent judiciary, claiming legitimacy due to the existence of a civil society. These two modes of governing became observable in almost purity during the lawyers' movement's protests, and thus provide an excellent case to discuss the theoretical framework of an analytics of rule as laid out earlier on.

Third, for the relationship of surveillance and mobile media in protests, this case highlights the importance of understanding them in a reciprocal relationship, where one answers to the state of the other. One might say that if the Musharraf regime at that time had the technologies and capacities to surveil mobile communications which became available to the Iranian regime during the 2009 post-election protests through the network equipment specialist *Nokia-Siemens*, the outcome of the protests in Pakistan and, much more, the accuracy of repression would have been different. Therefore, I suggest that the support that mobile media in protests provides is inversely proportional to the state's surveillance capacity: its finesse and granularity.

My general argument is that this case study highlights the necessity to connect both, the dissemination of mobile phones, its massification and roll-out, including the economic regime of its use, *and* the technological and operational status of surveillance systems during these protests, to understand the capacities of empowerment that mobile media may facilitate.

Pakistan and Musharraf

Some context is needed to look at this fight for an independent judiciary in Pakistan. Pakistan as a political entity emerged in 1947 during the process of partition when the Indian subcontinent gained independence from British rule. It is the home of roughly 170 million people, of which 97 per cent are Muslims. Formally, Pakistan is a federal democracy, but only nine years after its formation the first military coup took place.

> It was soft martial law. It does affect people, people do disappear, but it has never been a harsh martial law. Our dictators want to be popular also.
>
> Khan / Lahore

Since that time the country's rule has been dominated by dictatorship. Beginning in 1979, under the rule of dictator Zia, a deep islamisation of the country's structures and institutions took place, including the army. Even today the state and its structures are enmeshed with Islamic servants leading to a defensive position of secular groups (see Cohen 2004).

In 1999 general Pervez Musharraf, then Chief of Army Staff, in a bloodless coup d'etat overthrew the elected prime minister Navaz Sharif. He assumed power as Chief Executive of the army and on June 20th 2001 Musharraf appointed himself President of Pakistan, backed by the party PML-Q. He remained Head of the Army during his presidency and allied with the US in the 'war on terror'.[1]

His reign came to end in 2009, amongst other things, as a result of the protests described here. It would be an oversimplification to claim that the protests alone were able to push Musharraf from his presidency. "While there are a great many reasons why Musharraf was ultimately forced to resign [...] the sustained public protest of the educated elite class of attorneys and judges was clearly instrumental in that process"(Lyon 2010). The complex international political situation, the 'war on terror' in neighboring Afghanistan, and a diversity of national players need to be taken into account to understand the heterogeneous forces leading to his resignation. Still, it is safe to say that the protests had a strong impact with the restoration of the judiciary and an *elected* government in Pakistan.

The Dissemination of Mobile Phones in Pakistan

Mobile telephony was a liberalized market from the start in Pakistan, leading to five main companies offering services: Telenor, Ufone, Mobilink, Warid and Zong. The dissemination of mobile phones in Pakistan shows a significant rise during 2006 to 2008 (see figure 4.1). In 2008 every second Pakistani owned a mobile phone. "To describe the spread of mobile-phone usage between 2000 and 2007 as an explosion is an understatement, and such access became relatively common even among relatively poor peasants" (Lyon 2010). This coincides with the rise of the lawyers' movement: at the same moment in history when the larger part of Pakistan's society had only recently gained access to mobile telephony, a mass civil society movement was able to reinstate the Chief Justice and free the judiciary from a dictator's grasp.

In Pakistan, as in most countries, the largest share are pre-paid phones. Thus, although receiving SMS is free of charge, it remains unknown how many of these phones have credit to regularly send SMS or initiate a call. With regard to SMS

1 For a detailed discussion of Musharraf see Jones (2005).

Year	*Cum. Numbers of Users*	*New User in Current Year*
2001	300.000	300.000
2002	678.000	378.000
2003	1.532.280	854.280
2004	3.462.953	1.930.673
2005	7.826.273	4.363.321
2006	17.687.378	9.861.104
2007	39.973.474	22.286.096
2008	90.340.050	50.366.577

Table 4.1: Growth of mobile phone dissemination in Pakistan. Source: TechLahore.com

Year	*SMS in billion*
2006	8.70
2007	22.60
2008	61.00
2009	151.64
2010	175.39

Table 4.2: SMS sent in Pakistan 2006 - 2010. Source: propakistani.pk

though, the curve of numbers of sent SMS coincides well with the number of phones disseminated (see figure 4.2).

In 2006 approximately 8.7 billion SMS were sent through Pakistani networks; by 2008 this number had grown to approximately 61 billion.

But still, these numbers in no way reflect the number of phones used to send SMS and it is quite possible that a relatively small number of phones initiated a large share of SMS sending. Nonetheless, these numbers show that a countrywide boom of short messaging coincides with the first nationwide civil society movement. Asked about the costs of sending messages, all my interviewees in Pakistan considered SMS cheap.

> Actually in Pakistan text messages are very cheap. We bought unlimited packages. You could buy for 250 rupees 7000 text messages.
>
> Abbas, H. / Lahore

In 2008, 2,500 Pakistan rupees equaled approximately €2.50. Although this appears to be cheap at first glance, when confronted with the fact that more than half of Pakistan's population is illiterate, and with big varieties in gender and region, and that poverty is the country's most prevailing issue, SMS are clearly not available to large parts of the population. According to the 2009 Human Development Report of the United Nations Development Programme (UNDP), 60.3 per cent of Pakistanis live on less than $2 per day. In the UN's Human Development Report from 2009, Pakistan ranks only 141st of 177 ranked countries (see United Nations Development Programme 2009). Access to internet remains at 10% of the population "although the growing popularity of internet cafes makes it difficult to produce an accurate estimate of how many individuals have occasional access to the net" (Bolognani 2010, 404).

The Protest Movement

Some commentators understand the "Anti-Emergency Movement" as the first civil society movement in Pakistan (see Abbas and Jasam 2009). Sakham Khan from the Concerned Citizen of Pakistan (CCP) depicts one of the central demonstrations during the prostests very much in the tradition of U.S. civil society movements:

> At the first long march, when it took place, in July 2007, when we got to Islamabad, it was a feeling very much you would have in the western world, like the anti-war movements in the US, people would get together, happy, but upset. You saw in these demonstrations almost whole families.
>
> Khan / Lahore

The composition of the protest movement is heterogeneous to some extent, as Asad Jamal tells:

> The laywer's movement was heavily concentrated from the center Punjab, starting from the regions around Lahore, it was heavily concentrated in the urban areas. I think people from all kinds of backgrounds and affiliations were participating. When the movement matured, I guess that it turned somewhat against the Pakistan

> Peoples Party. The majority of the participants, it seems, they are inclined to support the centrist or right-leaning parties. Centrist because a lot of the people in the movement were against the PPP, because of Zadari or Bhutto. The media is either. There were a number of leftists, too
>
> Jamal / Lahore

Clearly urban centric, and led by the middle class, the Anti-Emergency Movement could encourage large parts of the population to engage in the struggles, although the backbone were the lawyers themselves and their organizations.

> If a Punjabi farmer can see in a Chief Justice's legal struggles a parallel to his own, if the bookstores in a city the size of Lahore can sell all of their copies of the constitution, and if the electoral success of political parties and the market share of television networks can hinge on their attitudes toward judicial independence, then the notion of the Supreme Court as merely a tool of elites cannot be the entire story. (Anonymous 2010, 1725)

Even political parties advocating the shari'a law in Pakistan backed the secular protest by and large. "The Jamaat-e-Islami was one of the parties most actively engaged with the lawyers' movement, despite the latter's aim of strengthening the secular legal system in the country." (Munir 2009) The common target was Musharraf and the goal of his resignation could channel and combine even opposed interests. Clearly, working classes involvement was limited and this has been "related to a general feeling of resignation about the recurrent use of martial law in the country, the fact that social justice issues (such as the rise of flour prices) would have been much more appealing to them, and also a widespread disenchantment about democratic rule" (Bolognani 2010, 403). Nonetheless it is safe to say that, at least from the phase of martial law onwards, heterogeneous forces joined with the single goal of an independent judiciary that eventually led to Musharraf's resignation. Maybe one of the reasons for success originated from its perception as being authentic: "The movement was not dominated by any single ethnic or sectarian group, thus giving it a truly national complexion. And it was not a facade for political party activism, although parties occasionally participated—especially at key events"(Munir 2009). Additionally, it is important to understand the role of the Chief Justice himself, who became an identification figure for the masses, what had become rare in Pakistan:

> I think for the first time in our history somebody stood up to a dictator and said "No, I am not doing this". That was the Chief Justice. That is why people across the country, regardless of their province, stood up on this. And they had somebody to follow. We don't have leaders like that, people only fight for their own interest.
>
> Zaman / Lahore

The independence of justice had a face and was embodied in purity by the Chief Justice. This certainly helped identify with the abstract dimension of law.

A brief note about the composition and economic standing of the Pakistan lawyers, as well as its organizational model and a description of two other groups, complete this introduction sequence before I turn to the chronology of the events.

The Pakistani Lawyers

As in any larger profession, different social strata are present within the thousands of lawyers in Pakistan, depending on were they work, in what position and what seniority they have. Thus "lawyers in Pakistan also represent the middle class and the lower middle class, while the rest are part of what is commonly and broadly called 'the elite', referring to the upper classes" (Bolognani 2011, 22).

> The lawyers have associations all around the country, so that helped. And then like maybe there are 10.000 lawyers in Lahore, that is impressive, and their dedication: they just went on and on, without getting jobs. It was their commitment that made other people follow.
>
> Zaman / Lahore

This commitment, it has been commented, would often find its origin, not so much in the interest of the idealistic good of independent judiciary, but much more in a material interest. As an independent judiciary strengthens the powers and position of the lawyers and increases their income, as opposed to dependency and good will from a dictator.

> Beneath the idealistic sheen of the lawyers' protests, the strategic maneuverings of elites were a prime driver of the movement and a key

> determinant of its success. First and foremost, the lawyers themselves were from an elite profession, and from a certain perspective their protests could be understood as a narrow defense of the professional interests of the bar. (Anonymous 2010, 1719)

This is truly the case, but the ones taking to the streets, engaging in risky operations during the protests, and as well lacking any income for an unforeseeable time were most certainly not, primarily, calculating their personal interests. "[M]uch of the impetus for the lawyers' protests came from the youngest and often the poorest members of the profession. These so-called 'common lawyers' tended to be more politically active than their elite counterparts." (Anonymous 2010, 1720)

> I was not in the position to pay the rent anymore, telephone was disconnected. It was a period of one and a half years. I had to take my children from school, I have not been able to pay their fees.
>
> Mohammad Azhar Siddique / Lahore

The other outstanding point is that hardly any other profession can claim that their professional interests, and the interest of the public, for the fair and due process of law, fall together.

The backbone of the lawyers was their high level of organization through bar councils and associations.[2] The protests did not emerge out of the blue and did

2 Zahid Shahab Ahmed gives a concise overview of the organizational structure that was imperative for the functionality of the lawyers' self organization. As I am not primarily interested in social movements structures, but focus on mobile media use, I take the liberty to quote in full length but in a footnote: "The 160 local bar councils and associations in Pakistan, with their 85,000 to 90,000 members, comprised the organizational backbone of the movement, which was led by a small group of lawyers headed by Aitzaz Ahsan. The movement's main decision-making body was the National Action Committee of Lawyers (NACL), whose main leaders were Ahsan, Ali Ahmad Kurd (president of the SCBA), Munir A. Malik (ex-president SCBA), Hamid Khan (member of the Pakistan Bar Council (PBC) and ex-president of SCBA) and Tariq Mehmood (retired justice). Most decisions were made in consultation with the Pakistan Bar Council (PBC) and the SCBA, whose leadership was represented in the NACL. The lawyers' nation-wide infrastructure and decentralized character (albeit with a centralized leadership) strengthened the movement's organization and communication, enhanced unity of effort, and made it very difficult for the regime to suppress the lawyers' activities. In addition, the NACL benefited from charismatic leadership. One lawyer from Lahore said that that 'the NACL member, especially Hamid Khan and Aitzaz Ahsan, successfully dominated the media'. But charismatic leadership cannot compensate for lack of organization, which is where the power of nonviolent resistance resides." (Ahmed 2010, 502-503)

not lack steering and hierarchy. The lawyers' ability to mobilize lawyers across the country made their protests especially effective. Nonetheless, without hundred thousands of mobilized non-professionals, even such an organizational structure as the bar associations would most likely not have been successful at such a scale.

This was by and large achieved with a very effective interplay of peer to peer mobilization via SMS, a functional and ad-hoc cross pollution of online media and mobile media, and a strong support by broadcast media, which was informed and updated as well by mobile and online media. This successful interplay is the central part of this case study, but two shorter passages are instrumental for an understanding of the protests: first, the important role that elite students played, as they primarily provided the means for an online dissemination, and second a brief chronicle of events is necessary.

Student Protests and LUMS

Students from elite universities played an important part in the protests, e.g. the Lahore University of Management Sciences (LUMS), who benefited from privileges that made their protest effective and safer at the same time: as in other countries, also in Pakistan university campuses are 'autonomous' areas guarded by university security personnel.

> They have a private campus and like many others they have their own security staff. Actually there is some tradition of this in many other countries as well but maybe not in the US. There is a tradition for autonomy for universities. The government is not going to send troops into the campus. And so, at LUMS it was the same way. The students were doing a lot of organising on campus, while the troops were not going there. But if they left the campus, than they could be picked up. Students were arrested and prosecuted in some cases.
>
> Theriot-Orr / Seattle

Additionally, they had good access to online capacities and were the digitally literate to function as proxies between online media and mobile media.

> During this state of emergency when every news sources was banned, the students organized a political movement using the Internet's social networking tools, such as emails, online petitions, blogs, the Facebook and text messages from cell phones. (Shaheen 2008, 143)

I interviewed students from LUMS during my stay in Lahore and it appears to me that the anti-emergency protests were their initiation to political protests. This explains the furore and energy they engaged with. Also they played a pivotal role as they came as a surprise. Amber Voha, who was present in Lahore during the protests as a journalist, writes: "If there was anywhere Musharraf might have expected tacit support, many thought, it was among these children, many of whose families are closely connected to his regime" (Vora 2007). But on the other hand: that law students are keen to enter their professional life under an independent judiciary seems rationale, unless they plan to go abroad to find a better professional perspective.

Additionally, in personal conversation with students it became clear that some understood their engagement in the protests as well as a revolt against their parents, who broadly did not participate, and thus deligitimized themselves in the eyes of their children.

The students, although not yet lawyers, belonged to a milieu that in different ways was connected to the lawyers.

> The student groups were largely law students, but not lawyers yet. The protest movement in Pakistan was much bigger than just lawyers. Although that was probably the majority: most people had some connection with the lawyers, either a friend, or parents. But there were also economics professors, anthropologists, various people from academia, as well as journalists
>
> Theriot-Orr / Seattle

The Pakistani students' connection to students abroad was instrumental in the distribution of information and funding of the protests.

> Pakistanis abroad participated a lot. During long march one we needed to raise funds and there were a lot of people who were sending us money through Paypal. One of my friends was managing this. And then there were people who were writing articles in newspapers abroad and they were sending articles to us to distribute.
>
> Salemi / Lahore

The special circumstances under which the students were working are expressed as "[w]here the Gates of LUMS start, Pakistan ends" (Bolognani 2011, 31). This gated community sphere is criticized by commentators as it would lead to mostly virtual protests that are considered non-instrumental: "the great majority of the student involvement was witnessed within the LUMS gates (and projected on the Web)" (Bolognani 2011, 32). But one can well understand that the students acted as providers of networked information, which played an important role in keeping communication between cities and to fellow students abroad, up to date. With this, the LUMS students were part of a functional division within the protests. Additionally, the organization of two long marches spurred their agency and thus they engaged in street protesting. I understand the students' contribution as a smart intermingling of online and offline protest.

> During long march one we had done a lot of work with the students in Karachi, Lahore and Islamabad. So there were some people who were actually doing an online reporting of what was happening during the long march. We usually sent a SMS and they would upload it on Twitter and Facebook and the internet immediately, on their blogs as well. Things like "we are so and so many now", "we are here now", "this just happened". So, people who could not join us could help us like this.
>
> Salemi / Lahore

Online communication did not reach the poorer parts of the protesters. This produced different informed milieus within the protest movement. In a heterogeneous and large protest movement, homogeneous strata of society stick together and use the tools and knowledge available to them in the sense that it is part of their agency and mentalities. Power relations prevail within movements.

Amongst the many publications the students provided, a flier by the Student Action Committee (SAC) (see figure 4.1) titled "Students demand a check on the establishment" directly addressed their claims: besides an independent judiciary, they addressed food issues as well as demanding free media.

After describing the general composition of the larger protest movement, I explained the heterogeneity of the lawyers themselves. This is often neglected, especially as large numbers of poorer lawyers hit the streets in an unanticipated way as they could well identify with the Chief Justice as a leader of integrity. Then I described that elite students backed the lawyers: first, they came as a surprise and second, they could operate safely from within the university gates and set up a cross media information network whose outreach was national and international. Next is a time-line of the events.

STUDENTS DEMAND A CHECK ON THE ESTABLISHMENT

With the removal of the pre 3rd November judiciary, there is, currently, a total lack of accountability of the **ESTABLISHMENT** which has:

- Shown no regard for our **fundamental rights**
- Proved highly inefficient in providing us with even the most **basic necessities** like **Flour**, **Electricity** and **Gas**
- Illegally **detained and tortured** countless fellow **students** and **citizens**
- Allowed armed pseudo-student bodies to control academic activities in major universities while **suppressing student unions**
- **Failed** to meet projected **economic growth** & development targets
- Failed to provide adequate **security** for our **lives** and **property**
- Installed a **puppet Judiciary**
- **Curbed the freedom of the media**
- Resorted to **false propaganda** to clarify its position
- No **credibility to hold free, fair and transparent elections**

- بنیادی حقوق کی پامالی
- آٹا بجلی اور گیس جیسی بنیادی ضروریات کی فراہمی میں مجرمانہ کوتاہی
- ہزاروں طلباء اور شہریوں کی غیر قانونی اور غیر اخلاقی حراست اور تشدد جیسے جرائم
- سٹوڈنٹ یونینز کو دبا کر، مسلح اور نیم سٹوڈنٹ عناصر کے ذریعے اہم تعلیمی اداروں پر آمرانہ تسلط
- معاشی ترقی کا جھوٹا پروپیگنڈا
- شہریوں کی جان و مال کی حفاظت میں مجرمانہ غفلت اور ناکامی
- ذاتی اقتدار کو طول دینے کے لیے کٹھ پتلی عدلیہ کا قیام
- میڈیا کی آزادی پر ناجائز پابندیاں اور حملے
- اپنے غلط کرتوتوں پر پردہ ڈالنے کے لیئے مسلسل جھوٹی مہم
- آزاد شفاف اور غیر جانبدار انتخابات کے تقاضوں کو پورا کرنے میں ناکامی

Gather on **Saturday, the 2nd of February** and stand loud and strong at **NassirBaagh at 2 pm** to peacefully voice our collective concerns.

The Student Action Committee

STUDENTS DEMAND A CHECK ON THE ESTABLISHMENT

Disclaimer: We represent students united against these injustices. We are not affiliated with any political leader or party.
Join us: sac.lahore@gmail.com
Update on activities: http://pakistanmartiallaw.blogspot.com http://nseofpakistan.blogspot.com http://fastrising.blogspot.com

Figure 4.1: Flier by the Student Action Committee, February 2008

A Brief Time-Line of the Events: First Sacking of the Chief Justice

Chief Justice Iftikhar Chaudhry can be considered the central figure throughout the events. He was appointed by Musharraf on 30 June 2005 as the highest judge of Pakistan. Up to his appointment, Chaudhry had not caused any trouble to the regime or made spectacular decisions. Though, since his appointment as Chief Justice he showed a different face. "There was a paradigmatic shift in the core function of Pakistan's Supreme Court when decisions started going against senior members of the executive." (Munir 2009) Amongst the many surprising projects, the launch within the court of a Human Rights Department to investigate cases of missing persons – people that had been detained incommunicado under the auspices of the Anti-Terrorism Act – was maybe the most provoking to the regime. "[F]rom 2006 to 2007, the Court required representatives from Pakistan's intelligence agencies to appear before it and account for certain missing persons that the government had allegedly detained in connection with the 'war on terror'." (Anonymous 2010, 1712) As Daud Munir puts it:

> In an astoundingly bold move, Chief Justice Chaudhry accepted a case involving 41 missing persons. The court upheld the right to due process for extra judicially detained individuals and ordered the state agencies to produce them in court. This was an extraordinary contra-authoritarian step in the judicial history of Pakistan. Chaudhry's activism generated immense respect among lawyers and the human rights community, and was widely reported in the burgeoning independent Pakistani media. Emboldened by the positive feedback, Chaudhry accepted another case involving over a hundred forced disappearances on March 8, 2007. (Munir 2009)

Additionally, he directly intervened into one of the most important privatization tasks of the regime: the selling of the biggest Pakistan steel mill to private investors. "In a landmark judgment, the Chaudhry Court annulled the sale agreement on the grounds that the deal had been done in 'indecent haste'." (Munir 2009)

Amongst the many accounts available, this one by lawyer Mohammad Siddique recounts the events of March 2007 from the perspective of a juridical activist who became a central node in the distribution of SMS to many lawyers across the country:

> And now the bad day came, the 9th of March 2007 when Chief Justice was thrown out from the court, from his premises. He was a custodian of the supreme court premises and 60 other judges. A deterior constitutional law has been used as a weapon. It is known as PCO. Musharraf got a PCO and urged the judges to take a oath under that PCO rather than under the constitution. There are 60 judges who were offered to take oath and some refused. That was the day when the movement started, the 9th of March. It was Friday, 12.30h, we were preparing ourselves for jummah prayer [friday prayer, O.L.], when we heard certain voices that there is gonna be some action against the Chief Justice. On that day in High Court Lahore was a proceeding against a famous general on the way. On that day these proceedings were withdrawn. And at about 2 or 3 we came to know that Chief Justice had been sacked and that they decided to take an action against the jurisdiction. We had developed a plan how to develop that movement. And on the 9th at night I had sent messages cross the country given them the news that Chief Justice has been sucked. To all lawyers. My mechanism are concerned with the heads of bar association, secretary, member of bar councils and the leaders of the bar.
>
> Mohammad Azhar Siddique / Lahore

When "Musharraf put an end to these investigations by deposing the Chief Justice Iftikhar Chaudhry on charges (not followed by official proceedings) of corruption" (Bolognani 2010, 402) and misconduct, Chaudhry refused to resign and thus was placed under house arrest by Musharraf. That the Chief Justice did not bow to the regime was unseen and unheard of for a long time in Pakistan, and it had a great impact, as lawyer Asad Jamal explains:

> And when he said "no", that took people to the streets. The patience of people had to come to the point were they had to come out, express their dissent, whatever political background they belonged to. And the media was ready to grap the opportunity. Media has had a explosion in recent years, a lot of investment came to the country, a lot of media grew, and the government gave them at least the freedom in establishing the channels. This was an instance when commercial media wanted to display what people wanted to watch. In the initial weeks of the movement in early 2007, the movement was on TV all the time
>
> Jamal / Lahore

To understand what this behavior of a high judge in Pakistan stands for, this anonymous writer from the *Harvard Law Review* explains:

> In Pakistan, where the military has dominated politics for sixty years and where generals demand a high degree of deference from legislative and judicial officials, the Chief Justice's refusal to resign represented a shockingly radical break from political and social norms and an extreme assertion of judicial independence. As one civil society activist put it, 'We Pakistanis are not used to people taking stands.' Within days, Chaudhry was transformed from a little regarded jurist into a folk hero. (Anonymous 2010, 1712)

When Chaudhry left his residence on March 13th, he was received by a police car to transport him to the court where he was to face charges. He refused to be transported in a police car leading to a harsh treatment by the security forces. "At this point, the security forces pulled him by the hair and forced him into the car. Their actions were captured on film and widely circulated on news channels and reported in the print media." (Munir 2009) Many commentators view this moment as the origin of the lawyers' movement: to see on news channels all around the country how the highest judge was mistreated.

The support for Chaudhry strengthened and developed pressure that eventually led to the unlikely event that on July 20th the "Supreme Court passed a landmark judgment annulling the presidential order [of the Chief Justice's sacking, O.L.] and restoring Chaudhry. This was the first judicial ruling in the country's history directly challenging the action of a military dictator" (Munir 2009).

Musharraf's rule was overturned by a court that had, up to that day, been generally in favor of Musharraf. This early phase of protest is recounted in detail by Mohammad Siddique, as well pointing to his personal achievements:

> On the 10th of March we had decided to launch a protest across the country. In Punjab alone are around 110 bar associations. Across Pakistan are about 240 bar associations. And later on we involved political workers, civil society, students, liberals, and again the mobile device played a pievotal role. We got numbers of activists, student leaders. We had sat out Thursday as a protest day for lawyers. After the restoration of the Chief Justice we stopped, but until then we walked on the roads every Thursday. And I myself with the help of NGOs and civil society organised a protest on Sundays as well. Even on a holiday, I with my family, all the people came out of their houses and protest in front of the houses of the supreme court judges

> at that time. This was my brainchild, I organised this in front of parliament in Islamabad and the court house.
>
> Mohammad Azhar Siddique / Lahore

The Second Sacking of CJ and the State of Emergency

Reinstated, Chaudhry continued with his independent juridical work, now additionally challenging Musharraf directly by preparing a court decision against Musharraf's reelection based on the question of Musharraf's eligibility to run for reelection as president while remaining in a military uniform. When the Pakistani Supreme Court was about to rule on the validity of Musharraf's reelection as president (see Anonymous 2010, 1716), Musharraf suspended the constitution and declared emergency rule.

> Under his directions, the chief justice and seven other judges were arrested. Musharraf replaced Chaudhry with Justice Abdul Hameed Dogar. Dogar promptly obliged by declaring Musharraf validly elected as president and by declaring valid Musharraf's National Reconciliation Ordinance, which provided immunity from prosecution to numerous corrupt public functionaries. (Iqbal 2009)

The other direct action initiated by Musharraf, as part of the state of emergency, was the ban of media. Thus, online media became crucial in the course of the movement. "Millions of Pakistanis were faced with a ban on about a dozen domestic and international TV news stations and curbs on newspapers due to the emergency imposed by the Pakistani government." (Shaheen 2008, 143) Due to the liberalized media scheme that made Musharraf a more popular dictator, now "30 privately owned channels were promptly taken off the air." Media was targeted also physically as "policemen raided the Islamabad offices of Aaj TV, an independent news channel, and attempted to confiscate the channel's equipment. The telephone lines of Pakistan's first independent news channel Geo TV were cut and their broadcasters were threatened with long jail terms" (Yusuf 2009, 6).

Additionally, demonstrations and public gatherings were banned, so that "thousands of lawyers and activists were beaten and arrested, including most of the leadership of the lawyers' movement. Students and other activists protested in defiance of military rule and faced brutal police crackdowns" (Anonymous 2010, 1715).

And as Muneer Malik, one of the lawyers leaders recounts in his detailed book "The Pakistan Lawyer's Movement. An unfinished agenda":[3]

> The imposition of Emergency was followed by a massive crackdown. Virtually the entire leadership of the bar was placed under arrest. Justice Tariq Mehmood was picked up from his residence and packed off to Sahiwal Jail where he was confined in a rat infested cell and had to sleep on the floor biting cold. (Malik 2008, 240)

Although martial law was invoked, it failed to mute and stop the protests. On a less visible level main media outlets continued to work:

> Even after the imposition of emergency, when all channels were closed, not being allowed to air any program, they worked, but the program was not relayed through the cable. You could watch via Internet, and the radio was there. I was actually interviewed by the Voice of America, it was the 2nd night when the judge had to vacate the house. I know many journalists who would stay up whole nights to get information, calling me and other people, to get information
>
> Jamal / Lahore

The martial law phase during the lawyers' movement actually fueled the protests and widened participation. What had been the outcome of a well organized profession now turned into weeks of nation-wide protests as "the lawyers began to cooperate more directly with opposition parties. Protests became much larger and more diverse as a result. In addition, new protest groups emerged and came to include not only secular urban elites, but also some poorer and more religious Pakistanis" (Anonymous 2010, 1715).

Pressure increased to the degree that Musharraf resigned as Chief Executive of the Army on November 2007 and on December 15th a 'light' version of the constitution was restored. General elections were held in February 2008. On March 24, the newly elected government released Chaudhry, his colleagues, and his family from incarceration. Musharraf resigned under pressure on August 18. Asif

3 This is a very detailed account from the view of a high judge that contextualizes the lawyer's movement within the bar association's inner political fights before and during the protest. In its detailed and dense description, including the names of hundreds of judges and lawyers, it provides an internal perspective on the bar's mechanisms and it is written almost like a diary, up to concise descriptions of every hour.

Ali Zardari, who promised to restore Chaudhry to office, was elected president on September 6 (see Iqbal 2009).

The new government took its time to reinstate Chaudhry and the other lawyers that had not bowed to Musharraf. The reasons are manifold and only by way of a deeper look into Pakistani politics, including the role of the different political parties and other public players, can this process be duly reconstructed. This is off limits of this analysis (for a more thorough analysis, see Abbas and Jasam (2009), Anonymous (2010), Ahmed (2010)).

The lawyers' and civil societies success story ends when they organized a second long march in March 2009. But "before the hundreds of thousands of protesters reached the capital, the prime minister announced the restoration of the deposed judges. Chief Justice Chaudhry once again resumed office on March 22, 2009" (Munir 2009).

After this brief description of the events, that focused largely on the phases from the first sacking of Chief Justice Chaudhry to the end of the state of emergency, the larger role of mobile, online and broadcast media can be taken into account.

SMS, Online and Broadcast Media

> In the early days of the emergency, SMS text messaging was lauded across the Pakistani blogosphere as the savior of communication in a time of crisis. (Yusuf 2009, 13)

The regime fought hard against broadcast media that was in support of the protests. First, through the state of emergency, they banned independent broadcast media altogether, but after lifting martial law, the pressure on the media continued.

Under his rule in March 2002, Musharraf had established the Pakistan Electronic Media Regulatory Authority (PEMRA), thus facilitating the opening of 56 private TV channels. Many local and national radio stations with private ownership emerged as well as a flourishing newspaper market.

This moderate freedom of the media came to an end in June 2007, when "the Pakistan Electronic Media Regulatory Authority (PEMRA) imposed strict regulations on live shows and forced those TV channels to drop their famous anchors" (Ahmed 2010, 505). But the harder broadcast media were hit by censorship, the more online media and mobile media kicked in. What is commonly referred to as civic media was invented for Pakistan in these weeks. "It was in this media vacuum that other alternatives began to flourish: the public realized that to fulfill its hunger for news in a time of political crisis, it had to participate in both the production and dissemination of information." (Yusuf 2009, 10) The production and dissemination produced new media structures, connecting broadcast media, online media, and mobile media (see also Shaheen 2008).

> As such, new media in Pakistan has helped old media survive. The result is a media amalgamation in which information is pushed to the public, promiscuously distributed across broadcast media, new media platforms, and various digital technologies to prevent being disrupted or corrupted by the authorities. Thanks to amateurs and activists, students and concerned civilians, a nugget of information can leap from local televised news broadcasts to YouTube to SMS text message to FM radio broadcasts to blog posts to international news reports – whatever it takes to go public. (Yusuf 2009, 35)

This amalgamation of very different media technologies had one prominent effect: by using all means available and not only using online media and blogs, larger parts of the populations, especially poorer parts without access to the internet, but equipped with a cell phone, were kept updated about the events and could be mobilized within short notice. "The coverage of the Emergency seemed mostly based on a circle of trust, created by an ad hoc virtual community who had, if not democracy, at least the rule of law as a common objective." (Bolognani 2010, 408) The students endorsed this conversion of online and mobile media and professionalized it to the extend that every minute updates would be delivered by SMS to online sources:

> There was a blog or a site. We text the contents of what is going on right now, e.g. during the long march, people started from one end of Pakistan going to the capital. People from Karachi usually text all their members and to the blog. There was number which you could text to and the site would automatically rearrange the text online. So that site gave the people in Lahore and Islamabad updates. It was a start up company running that site, a blog from a Doctor in Karachi and CN report. CN report was a start up from a LUMS doctor. You could MMS the pictures to CN report and the pictures showed up. The number was associated with the server

> there. They connected the server which was running from a GSM server. It was open posting. And this number was published. It was our source of information. Media was not broadcasting the news of the long march at that time. And before, during the emergency, channels were shutdown. So, we developed our own sources. E.g. if I text, then the messages online would appear like: Abeer Hamid texted this.... And my fellows see that and know it is a reliable source. You had to register if you wanted it to appear as a text from your number, otherwise it would appears as anonymous text. This was nation wide. All the people that were texting with the same keywords, that messages would all appear online as a stream.
>
> Abbas, H. / Lahore

But the students as well operated an infrastructure that would provide a dissemination of SMS:

> Another site was at LUMS, it was a mass messaging site. You text to a number there and that would be delivered to all the registered numbers there. It was a PC that is online 24h and from internet it has a gateway to SMS. If you write "save pak", then you join the group "save pak" and all your text will be delivered to the lists "save pak".
>
> Abbas, H. / Lahore

In addition, emails by the internet-connected upper classes "inevitably included synopses that were copied as SMS text messages and circulated well beyond cyberspace" (Yusuf 2009, 5). Information flow across social boundaries was initiated. This included Pakistanis in the diaspora who"created discussion groups on the social networking site Facebook" (Yusuf 2009, 10).

> Video recording and picture taking, besides SMS, was important. We were able to move images very fast into the web. Especially the students exploited this feature. The people overseas were constantly being updated, so someone studying overseas was able to move that information back and forth, that did put a lot of pressure on the media. That was to me the most phenomenal thing. I could take a picture and put it on Facebook while I was there. We had a CCP group on Facebook. The social networking is now connected to the phone.
>
> Khan / Lahore

The heterogeneity of the protest movement thus resonated within the diversity of media technologies available. And in return, the heterogeneity of the media technologies ensured that groups from almost all societal strata got into the information loop.

> In most cases, emails included short messages that were meant to be copied and further circulated via SMS text message. The parallel use of SMS text messages allowed activists to reach a wider audience while continuing to keep information about their monitoring activities restricted to trusted recipients. (Yusuf 2009, 29)

Within this rhizomatic media structure, the lawyer's kept circulating SMS in their inner organizational structure, to the extent that one nationwide inner communication circle existed within the protests.

Mohammad Siddique recounts numbers of SMS he sent personally on his private phone:

> I had then build up a mechanism to coordinate the lawyers across the country. I had sent a lot messages. In these days the quantity of messages was less than 10000 a month. But when the movement starts it jumped to 25000 a month.
>
> Mohammad Azhar Siddique / Lahore

But behind this capability was a tedious analog work. To be able to reach all the lawyers he had copied into his phone the phone numbers printed in the bar association directory:

> I put in all the numbers by hand. And normally when I am going to convey a message to them there are about 3000 to 4000 numbers which I have on my list to whom I have to send messages. Whether they are living in Karachi, Sibbi or NW certain districts, Shangla, Peshawar and so forth. All provinces.
>
> Mohammad Azhar Siddique / Lahore

This backbone of coordination by the lawyers "enhanced intra-movement communication and coordination. Most movement instructions originated from the NACL and were sent to the heads of the local bar councils via cell phone and SMS." From there "the local leaders, in turn, communicated with the local bar members" (Ahmed 2010, 505).

As for another active group, whose agency was restricted to Lahore, the Concerned Citizen of Pakistan, Hamid Zaman explains their main motivation to use SMS and how it was used:

> All our messages used to go out by mobile phone because a very small percentage of people have email and internet. Out of the people that were with us maybe 20 percent. And so the mobile phone was the most effective way of mobilizing, because you are always carrying a mobile and so I can send you in the morning a message for a demonstration in the evening and you would read it some time in the day. So all messages used to be send via mobile phones. Any direction and instructions that needed to be given. We had a CCP number, one central phone, so that any person who wanted to send back feedback would send it to that number, that number was monitored by a secretary all day. This could be used for people to send comments. Any feedback anybody wanted to give would go there.
>
> Zaman / Lahore

As SMS had become the main technology on the ground for updates and inner communication, it is because of its instant mobilization ability in tense situations:

> In Pakistan SMS were absolutely key. People would organize in an hour or two in a way that would be equivalent to that idea of flash mobs that is trendy. In Pakistan they would not call it flash mob, but they would have these text messages saying: "go to the chief justices' residence in fifteen minutes" and then people would be there.
>
> Theriot-Orr / Seattle

But SMS as well provided a function of gluing the protests together in many different ways, such as jokes about the regime that circulated on a massive scale, as Asad Jamal highlights:

> People were sending not only informations, but all kind of jokes against the military regime. I think 99% in English, even if it was Urdu, it was written in romans
>
> Jamal / Lahore

The students used their own SMS information circuit that would fit into their specific needs, as Ahmed Salemi recounts:

> At a meeting we had the idea we would float a paper and take phone numbers of all these people to join the protests so that we could contact them for future protests. We did that, it was very successful, it was covered by international media, because a lot of students showed up. That's were we four or five people wanted to do

> something, we developed a strategy and float a paper where we put numbers. I started feeding these numbers in my mobile so that we could communicate, sending SMS and exchange what is happening. Eventually we texted people and they came to the second protest as well, there we again circulated the paper and took the numbers and I got around 50 numbers the first time, the second time around 300 and it went up to 400 numbers of students. Not only we circulated text messages regarding meetings, but we also started to circulate breaking news, if anything would come up, if someone is getting arrested or the state pressure against the judges or whatever. People were really applauding it. They had the feeling and wanted to get the messages, so while being very busy in their studies they could actually track the media of what is going on.
>
> Salemi / Lahore

Asad Jamal gives a general description of texting activities, of using the phone to record audio and video and forwarding information.

> Messaging all the time. I started subscribing to special packages in order to spread the messages. And of course for voice calls. I was told, soon after the imposition of emergency, not to use my mobile phone, so that I am not caught. And then people filmed the events with their mobiles. And we were speaking to media people and people were sending multimedia stuff around the country to update them. We were also recording slogans and taking them back home. Everyone who could afford to do it, did it. We sent messages to lawyers, journalists, friends. I used only my numbers
>
> Jamal / Lahore

It seems that for Pakistan the mobile phone found its ubiquitous multi functionality on a mass scale during these times.

> In the age of instant IT communication, which is global and decentralized, not only could multiple narratives (complete with 'facts') be distributed in almost real time, but they could not be entirely stopped by anyone. The Pakistani diaspora was no doubt instrumental in ensuring that, regardless of the government's capacity to shut down the mobile-phone networks during a crisis (which occurred only on very few occasions, to be fair), there was simply no way to completely block the flow of information. (Lyon 2010)

Such a heavy endorsement of SMS and mobile media causes dependencies on these technologies. This puts the success of protest movement at risk and individual actors lose their agency, once the technology is out of their hands. Mohammad Siddique gives a picture, that shows the limited understanding of SMS for protests within the authorities, although the ban of SMS was announced:

> I had been able to manage a mobile in our jails. And I forwarded messages to my wife who forwared it to all the committees. And I remember the day when an attempt was made to ban SMS. At that the time minister of information specifically declared the agenda that they are going to ban SMS.
>
> Mohammad Azhar Siddique / Lahore

As SMS was the main information currency of the protests, one might expect that the regime would in some way suspend its use or even ban it as a whole. Why this only happened in rare cases remains, to some extent speculative, still, in the next section reports of surveillance attempts are discussed that provide further legitimation on the speculations as why the regime could not control this 'SMS uprising'.

State of Surveillance in Pakistan during the Lawyers' Movement and Anti-Surveillance Tactics

The overall importance of mobile media has been described at large. In chapter 5 I discuss how distributed action ,facilitated through decentralized, peer to peer mobile communication, can be understood as an antagonistic model to centralized surveillance. Thus, accounts of experienced surveillance are discussed in this section, that hint to the larger type of surveillance applied during the protests.

In the course of the protests, one event highlights the importance of mobile media. Here, the regime intervened decisively to reduce the impact of mobile media during the protests. It shows how a central figure can be muted with relative simplicity once identified. The ousted Chief Justice, from within his house, were he was arrested under emergency rule, "chose to address the nation via cellphone" (Yusuf 2009, 12). The trick to enhance a usually one-to-one media technology to a mass broadcast technology, shows the flexibility mobile media has once it is used beyond personal communication:

> In his talk, he called for mass protests against the government and the immediate restoration of the constitution. Justice Chaudhry placed a conference call to members of the Bar Association, who relayed his message via loudspeakers. That broadcast was intended to be further relayed by members of the crowd who had planned to simply hold their cellphones up to the loudspeakers to allow remote colleagues and concerned citizens to listen in on the address. (Yusuf 2009, 12)

This ad-hoc construction of a rather classical one-to-many media infrastructure can work if dissemination is relayed through many nodes, each amplifying to either other places, or to a loudspeaker. To bridge the limited multiplying functionality of voice calls via analog loudspeakers is the technical hack. Having many others relaying, by catching the loudspeaker's output with their phones, means a social distribution mechanism.

This distribution model failed, because the regime was well aware of what was going on and could thus intervene at the strategic points of multiplication: "most mobile phone services in Islamabad went down during Chaudhry's address, prompting suspicions that they had been jammed by the government" (Yusuf 2009, 12). This shows that the more concentration of power at one point (here: the Chief Justice's address to the nation), the easier to interrupt the communication.

Abeer Hamid, a student of telecommunications at that time, provides a smart analysis:

> During second long march, which lead to the restauration finally, they blocked the net in islamabad. And during a rally in Lahore, they blocked a particular cell, for two three hours. They can easily block from the MSC or BSC. Text messages can be blocked. It can be done from the core end, you can block the text or calls from a particular area. Without jamming the whole network, you can block text from a specific area. As well of specific content, but there they blocked all texting from that area. We observed that thing at that time.
>
> Abbas, H. / Lahore

These "sporadic efforts to cut telephone lines and jam cellphone networks were common" (Yusuf 2009, 12), focusing on focal points of protests, as "mobile connectivity at the Supreme Court, protest sites, and the homes of opposition politicians and lawyers who were placed under house arrest was jammed at different times" (Yusuf 2009, 12).

This is seconded by Hamid Zaman for Lahore:

> At other times what happened was that the entire phone system would be jammed for a few hours, we couldn't send messages or even call people. When something big would happen they didn't want the information to get out. I think several times for a few hours the networks were blocked
>
> Zaman / Lahore

I agree that "the government's attempts to jam cellphone networks during the emergency demonstrates that, much like television, cellphones had become an integral medium of information dissemination and community organizing across Pakistan" (Yusuf 2009, 12). The specific point for this analysis thus is not yet made: the large scale dissemination of cell phones and its use for means of protest was new at that time in Pakistan. The regime tried to suspend strategically important communication with limited means only: either by focusing on known leaders of the movement or jamming areas that had already proven to be of special interest:

> There were many efforts to smuggle cell phones into the house of the Chief Justice during his house arrest. And then they just set up phone jammers around his house. In that sense, there was active jamming.
>
> Theriot-Orr / Seattle

The Pakistan authorities at this time had not been prepared for the empowerment and increase in agency via mobile phones. I assume that no proper surveillance technologies had been in operation for mobile telephony, such as the FinFisher or NokiaSiemens surveillance equipment and services. Many of the interviewed affirm this. Asad Jamal, a central figure of the lawyers in Lahore during the times of emergency, estimates that:

> Surveillance vis-à-vis mobile phones. I don't know. I assume that people were being tracked, but I don't know if they were tracked because of their phone. And I know that a lot of people stopped using their phones. I called Mr. Hamid Khan after the imposition of martial law and he was never caught. He remaind underground for a long time. I am not too sure about this technological savvy of the intelligence agencies in our country, most of the time they rely on physical resources. Rather than using very sophisticated instruments. I am sure they have developed some knowledge by now
>
> Jamal / Lahore

Taking such an estimation as the base for consideration how to protect oneself, even basic measures would help, such as using code words:

> I decided not to send messages at some times. Sometimes we decided not to use the word "army", so we used a code word like "the strongest forces of this country" or the "only establishment"
>
> Jamal / Lahore

Another principle that helps, when direct tapping of phones is the prevalent mode of surveillance, is to relay communication via other people. This mechanism was used by the inner circle of lawyers:

> I know people who were being tapped. People were alert to that fact. People had certain phones that they were using for certain activities. Some of the lawyers leader were very alert to that fact, some even wouldn't carry phones, but people around them would. It was almost like you created circles around the leadership.
>
> Khan / Lahore

Tapping still remains a strong tool of classical surveillance techniques. A fact that the students realized, as Abeer Hamid recounts:

> For three months it was very hard. Many people were chased by the military intelligence in the university. They came to the director of the University and gave him a list: "these people are involved in the movement, don't let your campus be used, do not interfere with the military, it is for the national trust, you can not allow your students do this." Which was all due to the mobile phones they were tapping.
>
> Abbas, H. / Lahore

Nonetheless, tapping of phones requires a lot of man-power. It is not a machinic operation, such as data retention. For every phone tapped, somebody needs to listen in from time to time. Computational analysis of speech is unlikely to have been applied here. Even if the authorities only identified the phone devices that would concentrate on protest sites and produced lists of people based on device id's, this still would have been a rather manual operation.

Additionally, relational surveillance, the (re)-construction of social structures by analyzing meta data of communications and thus reproducing the social network in operation – this most powerful technique to identify larger groups – was not in the hands of the authorities then. With such an analysis, important nodes of (re-)distribution of information can be traced down and silenced, even when the protests are in the thousands. To suspend protests, this has to be applicable in almost real-time. These systems, which only started to appear from 2009 onwards, first publicly noticed in Iran, signify a whole new era of mobile and surveillance media.[4] The Pakistan lawyers could still operate in the pre-era of fully established mobile phone surveillance.

Jamming of the cell phone infrastructure is like using a sledgehammer. It is the most simple tool of force, as it cannot differentiate or discriminate within its area of operation. Either "on" or "off". Such methods can initiate irritation and fear:

> Once there was a very strong rumor about Musharraf and I remember that started because the telephones got jammed and nobody got through and by evening people met and said: something big is happening, maybe that's why they shut down the phones.
>
> Zaman / Lahore

While direct phone tapping and jamming of the whole network, or particular cells, have apparently been the means of the regime, the most basic safety-enhancing techniques by the protester have been the use of code words, not using the phone at very critical moments or by very important persons, and a general awareness. Additionally, protesters' assessment of the capabilities that the regime had to surveil was conservative: although Pakistan's secret and military agencies employ many servants, they were generally not conceived as being precisely smart in their operations. The US lawyer Devin Theriot-Orr compares two different 'cultures' of intelligence agencies (US and Pakistan):

> It is different in Pakistan, because the government is not very discrete when it comes to surveillance. In the West, people don't understand surveillance, in part because it is much more subtle.
>
> Theriot-Orr / Seattle

If relational surveillance is both powerful and far more commonly applied in Western societies, one can ask if these two modes, the 'sledgehammer' and the

4 The deployment of state of the art telecommunication surveillance systems rented or owned by governments has changed dramatically since. Almost all state has deployed a rented or bought commercial system, such as the products from FinFisher.

silent collection of relations amongst people, are expressions of two different principles of governing. I address this in the next section, where I try to relate the actions taken by the Pakistan regime in sovereign governing mode, and the actions taken by the protesters in governmental mode.

The Clash of Rule

I argued in section 2 that the framework of governmentality can be applicable to societies and regimes outside of 'the West'. This, I argue, is as well the case for Pakistan, an Islamic country, but to make these claims more comprehensive, some limits on the application of governmentality for Pakistan are helpful.

There are many fields and issues that are less well to see with a governmental lens: particularly the poorer regions, where close to slave labor like conditions prevail in agriculture or in the NWFF Provinces, which are tribal territories. I would find such an analysis difficult, because a main condition is hardly met there: the autonomy of the modern subject, invested with rights.

For governmentality, the modern state and the modern, autonomous subject are but different sides of the same coin. By separating the sphere of the state, comprised with its servants and institutions, and society, comprised of autonomous, free subjects invested with rights, a totality for governing had been established. Thus, the population could become both, the target and the means of governing. Free subjects are "individual or collective subjects who are faced with a field of possibilities in which several kinds of conduct, several ways of reacting and modes of behavior are available." In this design "face-to-face confrontation of power and freedom" is impossible as "mutually exclusive facts" (Foucault 2000f, 342). Being invested with agency and a set of possibilities and choices are the key ingredients of governmental rule. The urban elite and middle classes that were dominant within the lawyers' protests acted on the base of their own will, in the sense that the "governed are free in that they are actors, i.e. it is possible for them to act and think in a variety of ways, and sometimes in ways not foreseen by authorities"(Dean 2010, 21).

Such a predisposition cannot be switched on or off arbitrarily, but develops historically over time. To understand oneself as invested with possibilities of action is part of education, learning, and the challenge from one's social environment to act independently; to pursue one's own goals. These mentalities are assisted and further developed to find their own expressions with the dissemination of mobile phones. One of the dominant political technologies of governmentality during the lawyers' protests had been mobile media. To some extent mobile media perpetuate pastoral power, or better: pastoral power, a cornerstone of modern rule, can gain a high degree of operability through mobile media. Being able to communicate freely from the state often falls short with the loosening of ties to traditional bounds, such as family. The meaning of this for Pakistani or Indian society can hardly be overestimated: for example where the segregation of male and female teenagers is harsh, a whole new field of communication was established, where for the first time these teenagers can communicate freely from this harsh segregated regime with others. A teenager who took part in the interview I did with the alternative education project Ankur in Delhi describes this very vividly:

> You find 100 SMS in my phone, but you won't find ten in a girls' phone, because she deletes them. She doesn't want to read the message to anyone. Boys have less privacy. In our locality, LNJP, girls need only communication. The meaning of communication for boys is different to that for girls. Boys have different characters. Society thinks she is getting in contact with boys. You write a letter to someone, but a letter to a boy has a different meaning. Same with SMS. This is also a community problem. This is not a specific girls problem in a way, because there are differences within each community. She was using a phone for six months without her parents knowing this. This is not about hiding the telephone. The meaning of the world according to the parents is different to the girls that communicates via phone. There are different personalities. I can use my 2nd personality in chat and phone.
>
> Anonymous People from Ankur / Delhi

If "the meaning of the world according to the parents is different to the girls that communicates via phone" the power of mobile media is obvious: it makes possible the construction of new meanings and subjectivations because it allows communication beyond hegemonic boundaries.

But why is mobile media a new pastoral power? Because it facilitates expression of feelings, accounts of personal stories and generally instructs to jabber. Whatever the precise reasons, the fact can hardly be denied. So, if mobile media triggers a modified pastoral power within users who engage in intense mobile telephony, this happens largely irrespective of the cultural tradition, the religious background or the segregational degree of communities; it follows a transversal power vector, one might say. The process that integrates freely chosen personal communication into quotidian life at the same time stimulates governmentality as rule as the process of subjectivations initiated with mobile media, as well initiates a sense of independence and freedom: the possibility of choices. This produces a troubled tension between mobile media users and the powers of discipline, which targets the formation of "docile bodies", or sovereignty, which prohibits. Foucault's third pillar of power is still being unfolded today, in Western and non-Western societies, with specific outcomes and through recent political technologies: mobile media.

The economic regime of mobile phone communication helps to bring to light these new subjectivities. Most of the Indian subcontinent is penetrated with low cost phones and prepaid schemes (see as well 4). Devin Theriot-Orr gives an example for Lahore:

> This family who we were friends with there. It has a husband, wife and five kids. And they were in the Christian minority in Pakistan, which is very small, two percent of the population. Quite persecuted, marginalized, not having good jobs. And in that family every kid had a cellphone. They were in that category of working poor.
>
> Theriot-Orr / Seattle

And he refers to the general availability on the subcontinent:

> The other thing that is different to the US is that everybody in Pakistan and India has a cell phone. The rickshaw driver who earns 2 dollars a day has a cell phone because unlike in the US, incoming calls are free. There is no ongoing monthly contract. So you spent one time for getting a used phone for five dollars and then you have no running costs and text messages don't cost anything to receive. And sending them is quite cheap. Certainly people who are really poor are not going to have a cell phone, but even people who are barely the working poor will have access to cell phones which is very different to the US.
>
> Theriot-Orr / Seattle

Returning to the lawyers' movement or anti-emergency protests, the first thing to reconsider is that it were urban, well-educated and media savvy protagonists. Marta Bolognani states that "the effects of the interaction with new communication media by the educated classes both as consumers and as actors not only helped the protests, but for the very participants opened up ways of thinking, strategies of self-identification, self-projection" (Bolognani 2010, 408). Zahid Shahab Ahmed also identifies the impact on thinking and identification of the participant and civil society as a whole, as the "lawyers' movement was significant [...] for its impact on Pakistani civil society, which saw first-hand how organized grassroots activism could achieve results" (Ahmed 2010, 508). But it important to remember that civil society in Pakistan does not equal the population of Pakistan; it is much more comprised of the educated, urban middle class and elites that are affected by protests for a more democratic state, and as such entrenched in governing as part of civil society.

A further argument comes to hand when one takes into account that "many lawyers came to see themselves as participants in a narrative of lawyers' resistance to illegitimate state authority stretching all the way back to independence." And only one sentence later, "for many students, organizing against military rule had a profound impact on their political awareness and sense of self" (Anonymous 2010, 1723).

Seen from an analytics of government, it were different forms of governing, that are most likely always in tension and can be contradictiory by nature, which came in Pakistan to an overt, week long conflict, most drastically during the state of emergency. Not because during the times of emergency, public gatherings were forbidden and but still took place; this is more an effect of the forms of rule. From within the logic of government, sovereignty could no longer uphold its territory, and governmentality claimed hegemony within the modalities of rule. This is not meant in terms of literal territory but in terms of agency of rule: sovereignty, in the manifestation of martial law, was overruled by governmentality in the form of a rationality that understood the independence of the judiciary as a truth, embodied in the many lawyers, citizens and students. One might say that the civil agents of governing, the members of civil society, performed the outcome of a governmental program seeded by Musharraf's regime itself, by way of liberalization of the mobile phone sector and the Pakistani media economy. The freedom to self-technologies, to self-motivation, to choices and to an economically self managed life-style took its share in governing as the draconian martial and anti-terror laws were *not rationalizable*; the gap between the regime's ability to rationalize its rule and the societies' desires widened too far. As I wrote earlier: governmental management produces mindsets, rationalities and performative settings that relate the subjects to the state no longer as an exteriority of incalculability and fear, but as a totality that has to rule in accordance with the population's needs and desires. If a regime fails to acknowledge this and these desires are not met, a clash between the forms of governing develops. In the case of Pakistan the scene was an overt month long unrest.

Mobile Media as an Agent of Governmentality

If the argument unfolded has teeth, then one might conclude on this basis that mobile media in protests is not only a tool for organization, for mobilization, or a danger to protests as it facilitates surveillance, but increases a very specific kind of power: one that goes hand in hand with the individual invested with a the Western rationality of a "free will", that is someone whose rationalizable horizon of truth is in accordance with a specific idea of participation and of self-management. Mobile media increases the self-perception as a righteous protester. Its immersion with mentalities is more than just a functional or tool-like link. It increases a specific condition of subjectivity, rearranges it and sustains it. Mobile media empowers specific subjectivities in its performance. It empowers and assures a rationality that acts out of its own belief into its managed freedom and independence. And it is in this sense that one can say that mobile media is empowering. As a tool, this is always the case and an evident, almost self-fulfilling claim: yes, mobile media allows better communication, for sure. But this is not enough to understand when, where and how. My claim is that mobile media needs to rest on or needs to trigger a specific mindset, it needs to resonate within this mindset and that is, in terms of an analytics of government, the activated subjectivity; the citizen whose agency is processed and produced within a governmental program.

This argument critically focuses on mobile media, which now appears to be part of governmental rule although it facilitates the protest against a concrete manifestation of government. I agree that the imposition of emergency rule was "a clear instance of the diversion of the regime's energies to fighting civil society rather than focusing on important policy issues or confronting the growing threat of Islamist militancy in Pakistan" (Munir 2009), but there is more: evidently the regime lost its cause. Civil society 'took over government' in the wide sense of the Foucauldian framework. An activated citizenry found its technologies of governing; mobile media. Mobile media facilitates those that learn to jabber, that are immersed with ‘pastoral power 2.0’. The rationality, the truth, that needs to be a sustainable layer all the way through the protests, resonated and communicated by SMS and voice calls throughout the country’s urban areas, on which a discourse then motivated agency to perform. It was amplified by online media and received its echo from Pakistani communities abroad. The flow of signs, one might say, continuously and nationwide, facilitated by mobile media empowered governmental rationalities.

From here, it looks promising to take a look at distributed mass actions, which are supported with messaging and mobile phones; to understand how this operates within a different specific topology, following a different specific media logic than common centralized approaches of surveillance. This difference of logic, decentral and distributed vs. centralized, opens the possibilities of coordinated action without a classical top-down hierarchy. This action, I try to show, thus can circumvent centralized sovereignty to some degree.

5 Distributed Action and Centralized Surveillance

The importance of distributed SMS use during the lawyers' movement was the key narrative of the previous chapter. It enabled a specific kind of coordinated action amongst the activists. Interestingly, although coordinated, it was by and large still decentralized and distributed.

This brings in another topic: protest movements that intensely use mobile media for coordination purposes are following the logic of distributed action: although the media technologies as such, here GSM, is technically quite centralized, still, its use is independent from this. Such an action is close to bottom-up processes, which establish their structure in the course of emergence in an ad-hoc manner and change their structure constantly, as new nodes pop up whereas others become inactive and drop out. The social structure that emerges is temporarily visible – otherwise it is hidden. Its backbones are the social relations amongst partakers of the protests, manifested in the address books of each protester's cellphone.

On the other hand, surveillance technologies operate much more from a centralized standpoint and structure, trying to oversee the whole. For the surveilling parties, it is hard to estimate where the tactical outbreak of protests will happen, as the crowd has no center. Nonetheless, the social relations amongst the protesters might be made intelligible via surveillance technologies. But then, either a real time analysis of traffic or a focused tapping of the coordinator's phones needs to take place. But this has to be arranged beforehand, sometimes by allowance from a judge, depending on the laws and also depending on the reality of separation of powers, which tends to be blurred in many states.

In structural terms, two different media logics are operating at the same time during protests – the distributed protest logic and the centralized surveillance logic.

While these two logics operate at the same time, they still are unlikely to converge at the same time. In between the distributed actions and the catch-up of surveillance stands a tactical window of opportunities open for protests to emerge. Once this is identified by centralized surveillance the window closes again; but it might pop up elsewhere.

In Foucauldian terms, mobile media can be understood as crystallization points, where two modes of governing – governmental and sovereign – realize themselves. The strength of mobile media governmentality is its independence from top-down organizing and planning. It operates tactically, but cannot free itself from the strategic settings and matrix laid out by 'sovereign mobile media'. Thus, it is strong in ad-hoc presence, but weak in holding a position over time. Sovereign rule, once it mobilizes its agents to a location, interrupts the spontaneity of distributed action. But it needs time to do so - it mostly fails in case the protesters disappear fast enough. This indicates that different kinds of governing are organized in different modalities of time and space: mobility and deterritoriality here, immobility and territory there.

In some instances, where surveillance has caught up with governmental mobile media, the empowerment of agency is reversed into disempowerment. Nathan Freitas recalls this in his narration about the Republican National Convention protests in New York.

> In the US today the police have figured out that activist A would call activist B and they track their location. This happens through the RNC 2008. I have been working with people like that: don't call each other directly, set up an indirect number that bounces you to the other phone. This simple stuff is like really changing.
>
> Freitas / NYC

Distributed action and governmental mobile media are only as empowering as their practical counter-part cannot catch up with it: the surveillance technologies and their means to capture tactical media use. Avoiding direct device to device communication, using proxies that blur the social network, are strategies deployed by activists that have studied or experienced surveillance at work. On the other hand, such a strategic upgrade of activist technologies is, most of the time, incompatible with spontaneity and the idea of an open and fast growing mass of people. Thus, strategic media use has an effect on tactical media use. It depends on the concrete setting, on the circumstances and situation if a strategy that excludes spontaneity as a tactic can gain momentum.

Additionally, such a way of planning and deployment borrows elements of military thinking, like an arm's race, whereas protest culture, by and large, is inspired by solidarity and fun. Daniel Kahn-Gillmor addresses a general concern:

> I am not convinced of the emancipatory power of the global communication network as it is currently implemented. I think they are offering a lot of opportunities for centralized control and I would like to see things used as end to end communication but I am not convinced that that is what is happening and I am not convinced that in the end these tools weigh in a positive way over all
>
> Kahn Gillmor / NYC

This technical consideration, stated by a long time software developer and maintainer of many software packages, I agree, remains an issue. But notwithstanding the shortcomings of the contemporary technology deployed, its use can catch up with some of the shortcomings, when it is used in spontaneous and mostly catch-and-run tactical ways, as the following statements show.

Mobile Media Facilitates Collective Actions

What follows are prototypical examples distilled from the interviews, that very precisely describe the logic of dissemination for distributed action. Basically all interviewees expressed this practice, wherever they were located. The point is that a technology of individualization prominently serves collective actions. This contradiction is addressed in the second section of this chapter, 5.

Asad Jamal, a lawyer in Lahore, explains the common practice of SMS forwarding. He points to the fact that SMS that originated from him, went back to him, to the degree that he would resend his own messages.

> The messages have been forwarded as well, some came back to me. And it happened that in a hurry I did resend my own messages
>
> Jamal / Lahore

Someone like the lawyer Mohammad Azhar Siddique represents a node in the distributed network, that serves as a major distribution point. He copied the complete address book of the bar associations into his phone (hundreds of numbers) and arranged the database into subgroups of different places.

> When I sent a SMS to a member of a bar association, he would forward the message to his colleagues by just pressing forward. That's why the activists were on the street on 5th November.
>
> Mohammad Azhar Siddique / Lahore

As he was in the inner circle of the lawyers' organization himself, he was a special point of distribution, centralizing the social network to some degree. But nonetheless, after he sent the messages, there was no coordination where the receivers would send the messages to. The structure was undetermined and would only be established as a whole during the course of its emergence; in different materializations each time.

Activists from other countries express the same dissemination practices, leading again to the same distributed mode of action.

In Delhi, Madhuresh Kumar explains, the idea that SMS should be distributed further is sometimes expressed in the messages themselves.

> In Delhi for our own solidarity group events, for press conferences, demonstrations, we regularly use SMS as a medium to mobilize or even remind people, email as well, but every day when we have an event we send 100 to 200 messages to people to remind them. The text is a reminder. We send it with private phones. People also forward the messages to other people. We also write "please forward"
>
> Kumar / Delhi

Vijahan from Delhi reports about the efficiency of short-term mobilizations of media and how he organized his phone:

> The last phone I was using, there you would find eight different groups, one is called "Delhi Media", if there is an action where we only want to inform the media, for example in front of the prime ministers house for the Bhophal demonstrations. 12h was the action, 11h I sent out the SMS. The key photographers come. That's a way to get rid of the police. By email there is always a way it gets public, and the police then will do everything to prevent the action from happening. So then, it is much better to wait till the last moment, like one hour before and sent alerts to the media. It will be a quick action, maybe 20 minutes. In that kind of situation it is extremly efficient.
>
> Vijayan / Delhi

For Oaxaca in Mexico, Blax recounts the dissemination practice. Adding that one SMS then as well is passed over to others nearby, by reading them out aloud.

> In 2006 it was very clear, people came to a meeting, or a barricade or a camp by text messages. You can also send to other people that are in direct contact with other people. It is the massive use by the people. One person receives it and tells others about it. Like a radio
>
> Blax / Oaxaca (Mx)

In Seoul the tactic was integrated into a strategy: knowing that coordinating a meeting point beforehand would increase the possibility of police presence, only on the fly a meeting point for action would be negotiated, when the protesters were already moving.

> I explain how mobile media was used in the protest: it was used to gather people. Textmessages on cellphones were used. We call cell phones hand phones. We made appointments to gather, there were no points to meet beforehand. People picked certain locations in advance, but due to police actions, these locations were changed. People exchanged informations about blockages and went to different places, hard to spot fastly for the police
>
> Two Korean Activists

Ederic Eder recounts how he remembers the first large protests, which were by and large organized by SMS, the EDSA2[1] protests in Manila in 2001.

> My personal experience during EDSA2 was, I was going home and suddenly received a forwarded message that said that those who were then acting as impeachement judges, were having to open an envelope that people believed was containing evidence against Estrada [the president, O.L.]. The text message was calling on people. That was in the night of January 16th 2001. And then I received that message many times. When people were already gathering at the EDSA you could see everyone was texting, everyone holding mobile phones. And there were times when the network was down due to the volume of text messages sent. So, those are mostly calls, slogans, and then people were updating their friends, "where are you?" and "let's go!" So invitations to join the rally and even jokes. People love jokes, even now. There were a lot of Erab [nickname of the president, O.L.] jokes then. I started to send text messages, saying: "Erab will be the first president to go down by text messages. Congratulations to all of us"
>
> Eder / Manila

1 EDSA, Epifanio de los Santos Avenue, is the name of the longest highway in Manila.

Al Alegre, who works for the Foundation of Media Alternatives in Manila, emphasizes the role of mobile media in EDSA2, but is a bit more cautious:

> In EDSA2, because it was very much urban based and a strong participation of the youth, strong penetration of these sectors, the young and urban, SMS was used to coordinate a lot of stuff. I really would not go as far as admitting the hype that cellphones stopped the president, but it was part of the repertoire of actions how technology was used, to fuel the social movements.
>
> Alegre / Manila

Vincente Rafael discusses the 'power of txting' in regard to the Philippines. He, too, comes to the conclusion that the circulation itself is the key ingredient of 'txt power':

> The power of texting has less to do with the capacity to elicit interpretation and stir public debate than it does with compelling others to keep messages in circulation. Receiving a message, one responds by repeating it. The message is forwarded to others who are expected to do the same. In this way, the message returns, mechanically augmented but semantically unaltered. They crowd one's phone mailbox just as those who believed in the truth of the call they received crowded the streets of Metro Manila. (Rafael 2003, 409)

Friends send messages to friends, who then send messages to friends. Social networks emerge and manifest themselves in body presence over the use of address books. Thus, redundancies exists, messages are received many times. But no address book is identical to the other. Each node can contribute to the growth of the protests.

At the 2008 G8 summit in Japan, activists used both a distributed dissemination model, and a centralized point, which would resend the news:

> Last year we had the G8 summit and a big demonstration. At that time, SMS and email by mobile phone was very useful: which direction to go, in real time, where the police is and those things. For the event communication it was very helpful. Everybody had access to the central office and then the central office compiled the information and spread it.
>
> Yasuda / Tokyo

Returning to the protests against the Republican National Convention 2008 in New York, a sophisticated messaging system was set up, called "TXTmob", which would distribute messages according to subscriptions.

> The communications collective had their own team of scouts, which was actually a fairly common tactic for them. They would have scouts with radios to communicate with each other and phones to talk to a main dispatch who would then use TXTmob to send messages out. And actually I am not sure if they used TXTmob or Twitter, I think they used TXTmob, because they were lots of groups so they divided the city up into regions and also various sorts of tactics. So there were messages based on regions and tactics, you would not have to subscribe to everything.
>
> Goldberg / NYC

While this is again a centralized approach, it goes even further as it divides the communication into geographic sectors of the protest. This approach then faced harsh repression and the members of the communication collective were charged with felony: they were too exposed and too central.

> During the RNC, the communications collective there got raided and charged with felony and incitement to a riot. Just as a side note. I think the only thing that they were doing was actually TXTmob.
>
> Goldberg / NYC

Layers of Distributed Action

Without tapping as part of an ongoing investigation (see 8), or disturbing the cells, or blocking traffic (see 7), surveillance only be always late vis-à-vis the dynamics of distributed actions coordinated by social networks of mobile phone address books. To block the net is possible, once access to the cells is at hands, but this as well means a disruption of any other cell phone communications: that of economy as well as that of the police itself.

There is an advance of distributed structures to let protest emerge at some unknown locality. Structurally, surveillance is late in detecting distributed action. These protest structures are only powerful for some amount of time and only if they rearrange and reemerge somewhere else again. Coming and going, they fluctuate.

Mobile media is therefore contributing to an unplanned emergence of structures, which manifests itself temporarily by way of body presence on the streets, and might either be halted and shutdown by authorities if it stays in the light too long; or it may disappear fast enough, only to appear again elsewhere.

Daniel Kahn-Gillmor reports from New York about a planned protest action, where surveillance was most likely involved to shutdown the protests once they occurred:

> In 2003 I was protesting one of the Carlyle's group offices in NYC. They had a troop out to meet us like we were gonna be, uhm, I don't know, there was a handful of people who were about to block the doors of that building, but that was the most extreme, the vast majority of people had banners and were gonna go to work later in the morning. And they arrested all of us. There was no way that with that level of response that there was not surveillance involved in the planning and organization of the meeting to show up with the group that they had for a totally unannounced action. Something had gone on, whether this was a disproportional response based on "let's test out some tactics before the RNC comes in town next year", or whether they thought that we were much more organized than we were. A lot of us were locked up and held for the day without reason
>
> Kahn Gillmor / NYC

The limits of collective actions are determined by the degree they are planned, one might deduce. But there are other examples, as Legume reports.

> For secret things, like actions where we need to surprise, we don't talk about that on the cell phones, or use SMS and not even in the mailing lists, because we think even we have a non-commercial mail service, most of the activists are using windows which is not a secure OS.
>
> Legume / São Paulo

A self-limitation in the ability to communicate can secure the success of protests that rely on the effect of a surprise.

Mobile Protest Media Can De-Individualize

It is interesting how the individualizing powers of mobile media are countered during protests. The device, its personal use, its attachment to one body is not changed as such. But the trajectory to individualize is countered by the dynamics of distributed action: my argument is that while it does not specifically facilitate collective actions, it does not disrupt de-individualization processes in the performance of dissent. The trajectory to individualize is overwritten by the manifestations of bodies on the streets.

In all instances, for the protagonists involved such a partaking goes beyond only a temporal appearance within a mass of protesters, as the experience to be part of a collective action that is not orchestrated from above empowers individuals in their agency of performing dissent. This counter-practice leaves traces of counter-knowledge that can be re-activated. Such an inscription of distributed action into the bodies of the partakers, can be understood as an alteration of quantity (the masses on the streets) into quality (the repeated experience on the individual level). Processes of subjectivations are triggered by repeated experiences of dissenting agency. Mobile protest media then is a political technology of empowerment, on the level of performance of dissenting practices, not only on the level of independence from the state or loosening traditional ties for the individual. As the repetition of performed dissent and movements sediments into the body, the subjectivity encounters counter-truths. While the performance of dissent faces a technology of individualization, the physical layer of protests overwrites this trajectory during the course the actions take.

This points to a further aspect: the co-presence of the mobile and the body. Where the tele-connectedness enables a diffuse coordination of bottom-up processes over distance, the presence of the body materializes this tele-connectedness on the physical layer. As the two layers are discriminated (signs and material), they supplement each other. What the body perceives on the street reaches other bodies through the address book and centralized infrastructure. And what mobile media targets as an individual is a fragment within the collective performance of dissent. The centralization of infrastructure does therefore not interfere with the distributed action. Each layer's *modality* is independent from the other.

In the next chapter, I shall return to the discussion of the apparatuses of security and the ways to limit governmentality and individual agency by what one may call *neo-sovereignty*. With this, a differentiation of governmental rule by mobile media is envisioned that bridges the field of surveillance.

Part Two: The Liberal Paradox

6 Surveillance

> People that are being wiretapped have no way to ask the police about it. Right now as the informations are in digital, the telephone companies or internet provider are accessed by the police. The wiretapped person will never be informed about it. The first time will be in court. There are very vague information about how many wiretappings and what for.
>
> Yasuda / Tokyo

For most of the interviewees, surveillance is a targeted operation with the aim to repress and mute, to silence and to produce fear. It can be a prerequisite for investigation, or even direct repression. It can be an operation over a long time gathering informations about oppositional activities.

> Surveillance and repression walk hand in hand and that leads to fear and paranoia. The paranoia of being followed all the time.
>
> M / São Paulo

In most cases, it is seen in a direct relation with the activities that activists are involved with. Many accounts on cell phone wiretapping were delivered, all point to a targeted and also expensive operation, one which cannot be done on the whole population.

> Surveillance is about controlling personal information, monitoring it, the metaphor of the panopticon: you might be observed but you never know, so you internalize the surveillance. All the time paranoid, because you might be monitored all the time, that is probably the worst aspect of it. The idea that you have to control yourself on what you do in terms of how nasty that could be to the people that have the power to stop you doing it. And there is another stage, we have encryption, anonymity proxies etc, but the bigger enemy might be paranoia, which is a meta level on self monitoring, how much you have to monitor yourself about the others not to monitor

you. There is a risk of blocking the necessary flow of information here.

Anonymous2 / Madrid

Targeted, specific surveillance, ad-hoc and tactical, needs to be differentiated from the strategic surveillance schemes, which target up to the whole population by means of massive interception, like data retention schemes.

We don't write anything about politics in SMS.

Non Collective / Manila

Of course, there is a connection: once an event is triggered from mass surveillance, the surveillance infrastructure can be queried for more information. Targeted surveillance and strategic surveillance become harder and harder to separate, as one supplements the other.

This chapter is about strategic surveillance, as conceptualized and discussed in surveillance studies; most prominently the field of surveillance studies from the perspective of social sciences. In the next chapter, a deeper discussion of data retention and other strategic political technologies takes place. Therefore, this chapter's idea is to introduce the larger subject of surveillance, while at the same time comparing to the next chapter, the difference to my framing of surveillance.

The main body of literature on surveillance is about the big picture, thus many scholars speak of a 'surveillance society'. A term that I do not find particularly useful, because it totalizes surveillance. I also do not find the divide between commercial and state surveillance helpful, as the amalgamation of public and commercial spheres makes it nearly impossible to separate the two; except for analytical reasons. The outsourcing of 'security services' and IT to private companies is a common governance task in most countries. Maybe an effect of deregulation, it also shows the operational or modal structure of securitization.[1] A fitting example is the outsourcing of tasks of surveillance in Germany's Lower-Saxony, where police hired a company to send silent SMS (or 'pings') to suspect's phones.[2]

---

1 On the development of the private 'security' sector in the US, see Marquis (2003). On the concept of securitization, see the next chapter.

2 "Die niedersächsische Polizei lässt Verdächtige per 'stiller SMS' überwachen. So weit, so bekannt. Neu ist allerdings, dass die Behörden eine private Firma für diese Mobil-

> Activism nowadays is becoming difficult, there is state surveillance, and private. You can never tell what will happen to you, everytime you can be beaten up by the police. Or the government thinks you are talking too much on that thing, they can treat you badly. Things are like that, because we are passing a time, it is not fully democratic. And at the same time we are taking our chances
>
> Swapan / Dhaka

What is Surveillance?

I refer to David Lyon's standard work on the "Surveillance Society" (2001) for a collection of aspects of surveillance: David Lyon in his introduction to "Surveillance Society" suggests that "all societies that are dependent on communication and information technologies for administrative and control processes are surveillance societies" (Lyon 2001, 1). Surveillance societies, according to this concept, are the other side of the coin of information societies: surveillance and information are mutually inclusive. Surveillance, he continues a page later while discussing digitalization, "is any collection and processing of personal data, whether identifiable or not, for the purposes of influencing or managing those whose data have been garnered." Surveillance is part of a managerial program: not only about collecting data, but in a managerial way to use this data on programs and action plans on the surveilled. Lyon states that "surveillance capacities are used to sort and shift populations, to categorize and to classify, to enhance the life chances of some and to retard those of others" (2001, 4). Thus, surveillance is a discriminating political technology that consists of both care and control, and "is now a generalized social phenomenon in the sense that institutionalized monitoring is routinely practised by a range of agencies including, but as well going well beyond, the state" (2001, 30). Lyon refers particularly to risk management:

> Everyday governance in contemporary societies tilts towards the management of risks and especially risks associated with obtaining com-

funküberwachung engagiert haben", (It is well known that Lower Saxony's police does surveillance on suspects with the use of silent SMS. But it is a new situation that authorities have hired a private company to operate this mobile phone surveillance. [Translation O.L.]), Süddeutsche Zeitung, Mirjam Hauck, 27 February 2012, http://www.sueddeutsche.de/digital/ueberwachung-per-stiller-sms-niedersachsen-jagt-verbrecher-mittels-eines-privaten-dienstleisters-1.1295001 (accessed 4 March 2012). About 'Silent SMS' see section 8.

> pliance or containing threatening behavior. Surveillance is the means whereby knowledge is produced for administering populations in relation to risk. [...] In a quest for security, all institutions in whatever sector seek to minimize risk by finding out as much as possible about as many factors as possible. (2001, 6)

Finally, Lyon emphasizes that surveillance is enmeshed in a "social orchestration" of compliance and active, productive engagement with surveillance by large parts of the population as being in the information loop is a necessity for social participation (2001, 7).

What is convincing in Lyon's account is the normality of surveillance as an operation to govern. The notion of risk management of populations[3] resonates with the discussion of data retention in chapter 7. On the other hand, if surveillance is a ubiquitous means, omnipresent in everyday life and central to governance, it appears problematic to draw boundaries and mark the specificities of surveillance. One criteria, Lyon suggests, is the digital double:

> The problem of disappearing bodies is thus crucial to understanding surveillance societies. It shows how surveillance systems have arisen in an attempt to compensate for the disembodiment of many social relationships. [...] contemporary surveillance practices tend overwhelmingly to base themselves on abstractions rather than on embodied persons. (2001, 26)

This hint towards the abstraction of surveillance with the production of digital doubles and thus the disembodiment of social relations I will explore in the next chapter. It is here, that the powers of digital surveillance are at work. In the multiple re-organizations of data entries and further the production of their relations; no physical, embodied phenomena play a role here. A new realm has developed: that of the *digital doppelgänger* and its very own algorithmic relations.

Four Strands of Surveillance Theory

Lyon (in accord with large parts of the literature on surveillance) suggests as a scheme of a theory typology four strands:

---

3 On the 'risk society', see Beck (1992).

> Four strands of surveillance theory may be distinguished, each profoundly connected with classic conceptions of modernity. [...] Surveillance may be understood in relation to the nation state, to bureaucracy, to what might be termed technological logic [...] and to political economy. (2001, 109)

These categorizations are common macro sociological types. They mirror a problem that I see in surveillance studies by and large: they separate a sphere that is interconnected and relational and suggest that each by its own have an explanatory capacity. In contrast, my intent is to understand surveillance in relation to an analytics of governing that is neither state centric, nor of pure economic origin, nor technology centered. Bureaucracy is a part of it, as well all other heterogeneous factors.

By dividing spheres which are enmeshed in reality, analytical separation disrupts an analysis of *how* they work; of what rationalities they entail, what truths they produce. In short: the effects of surveillance conglomerates are not foreseeable. Governing is always a risky operations per se. A typology as suggested by Lyon (which only stands for the large body of work within surveillance studies) reproduces many assumptions which frame surveillance within a discourse of modernity, which itself is a product of modernity. Here I see a problem with many surveillance studies approaches. Modernity, as so often, is not very enlightening as an analytical term. I do not see what the explanatory gain is when surveillance is described as a facet of modernity: "magnification of surveillance capacities simply is one facet of modernity" (2001, 31).

And even more, in most surveillance studies literature I miss a genealogical account of subjectivity which enfolds and develops within government. Power, it seems, in most surveillance studies is external to subjects. Some ideas that subjects themselves are part of power and governing echo those notions of individuals who comply with surveillance, for example to take part in social life. But it is exactly here where most surveillance studies literature stops, at this crucial border that is itself a product of an episteme of modernity: the idea of the autonomous subject, static, independent, which at most faces problems by discipling surveillance in the pursuit of its life. It is not conceived as both an effect and a site of production of power.

Allmer (2011) has chosen a different approach. He discusses the main body of literature on surveillance by way of differentiating between a panoptic type and a non-panoptic type. I do not share his assumption that surveillance studies with a panoptic fringe understands surveillance as 'negative', whereas others see it 'neutral'. Foucault's genealogical work on the discipline society is not a 'negative' account. Much more, it is an analysis of technologies of power.

What follows is a brief account of the surveillance studies discourse, informed by the first wave of reflection that took place most prominently in the journal *Surveillance & Society*.[4]

Surveillance Studies History

Surveillance has received much attention in academic research lately. Sociologists, criminologists, historians, geographers, media scholars and other researchers have tried to shed light on it. Surveillance studies emerged as an interdisciplinary research field that has produced its own discourse on surveillance in the last decade. But only recently have the implications and theoretical framings of the discourse around surveillance begun to be discussed.[5] The main strands of this debate are focused around how to conceptualize changes of workplace, the logical place of technology in surveillance, questions regarding post-9/11 governance in relation to surveillance, and lately a rising demand for local accounts of surveillance.

In his historical reconstruction of the academic discourse on the 'surveillance society', David Murakami Wood (2009) focuses on a historical dynamic and then problematizes a prevailing lack of contextualization and localization within this discourse.

> In order to tackle these questions the history of the term *surveillance society*, both temporally and spatially, is examined in the four periods of its development: the late 1960s and early 1970s; the mid-1980s to early 1990s; from the mid-1990s until 2001; and from 2001 until now. (Wood 2009, 181)

Paraphrasing Wood, the first period of conceptualizing the surveillance society is characterized by, inter alia, the end of colonialism, the Vietnam war and

4 It is an open access journal, see http://www.surveillance-and-society.org/.

5 See Kirstie Ball's and Kevin D. Haggerty's editorial "Doing Surveillance Studies" (2005).

the beginning of the crisis of Fordist capitalism. In this dominantly US-centric view, one of the most prominent surveillance dispositifs initiated was a new central database. According to Wood, it is from here that surveillance studies took one of its impulses. The fear of misuse by several agencies generated a "fear of the state" (2009, 182), which reached its height at the Watergate affair.

From 1984 to the early 1990s the emerging neo-liberal regime of Thatcherism and Reagonomics is seen as a driver for intensified surveillance at workplaces, as finance and consumer-management coupled with the miniaturization and mass deployment of computing powers. Wood references Gary T. Marx as the initiator of the "Big Brother" figure's transposition from the famous novel (Orwell 1949) to research through his article "The Surveillance Society: The Threat of 1984-style Techniques" (1985).[6]

The Panopticon and the Database

The other powerful term that has accompanied thinking about the 'surveillance society' and which still prevails today, is Foucault's *Panopticon*. According to Wood, scholars such as Oscar Gandy referred heavily on Foucault's late 18th century dispositif concept while analyzing the operations of databases. "Scholars struggled with concepts, however, mainly those of how to theorize the database." (Wood 2009, 184)

In 1988, Australian IT consultant Roger Clarke introduced the term *dataveillance*. It is noteworthy that Clarke, educated in computer sciences, does refer to Foucauldian concepts. Clarke clearly distinguishes between a systemic border of dataveillance and the bigger scope of surveillance. Although at core this is a technology-centric approach, the reduction to systemic aspects of dataveillance has its benefits.

Clarke defines dataveillance as "the systematic use of personal data systems in the investigations or monitoring of the actions or communications of one or more persons" (1988, 499). He then differentiates personal and mass dataveillance. When suspicion has arisen, personal dataveillance becomes operational. He concludes by saying "effective controls over the new and emerging techniques of dataveillance will not be possible" (1988, 511).

6 However, Marks (2005) found the earliest reference in a study by James Rule called "Private Lives and Public Surveillance".

Wood's third epoch of conceptualizing the 'surveillance society' spans from the early 1990s to 9/11 and is primarily concerned with the mass deployment of CCTV-Systems in huge urban aggregations. A shift from the US to the UK notably accompanies this movement. Clive Norris and others have discussed the emergence of CCTV in public and private places (Norris and Armstrong 1999). Early traces of this emergence can be identified with the shift that the military industrial complex took at the end of the Cold War by "seeking to find new civilian markets and exploit governments in the risk society" (Wood 2009, 186). The immense popularity of CCTV as a field of research in surveillance studies, I suggest, might be an echo from the CCTV's symbolic value: the massive deployment of highly visible technologies in the urban scape which easily resonates well with the figure of the *Big Brother*.

Foucauldian concepts of discipling the subject have been connected to CCTV, because CCTV works in both directions: cameras are connected to further infrastructure, for a processing of images, but the other direction is equally evident: to induce normative and compliant behavior into subjectivities and to reduce deviance.

The last (and current) phase Wood identifies, starts with the 9/11 terror attacks. And it is here that he starts to question a global approach of surveillance studies.[7]

> Lyon, Ogura and Mattelart have been clear on the idea that there is a 'globalization' of surveillance. And certainly when societies other than the places where Surveillance Studies academics are concentrated are examined, one can see similar developments. But is this as simple as Anglo-American conceptions spreading to other nations? (2009, 187)

The impact of a globalization of surveillance thus demands, contrary to intuition, a precise framing of the analysis carried out. Otherwise, this is how I understand Wood, mostly Western concepts are reproduced globally.

7 For a recent history of the process how surveillance became a globalized governance scheme, see Mattelart (2010).

The Medium is Surveillance

It is in this tension of the question of locality and globalization where I suggest that a perspective on surveillance via mobile media as a political rationality is enlightening. The rather empty term 'surveillance society' can be avoided by way of a recurrence to the liberal paradox – the provision of freedom and its excess in control – to understand that surveillance is not simply an effect of modernity, but can be seen as a historical development that is an immanent part of liberal rule. Mobile media as an individualizing political technology bridges the gap between the local and the global as a medium for the production of the individual. And as such, the individual as a Western concept is exported globally as a raw model and is reconfigured according to local specifities. By way of this export, mobile surveillance media as a material-semiotic conglomerate, couples and attaches to this trajectory. Mobile media is therefore also a global carrier of securitization. Mobile media globalizes surveillances. To paraphrase and alter McLuhan (2003): the medium is the surveillance.

Big Brother's Effects on Surveillance Studies

Additionally, Western concepts are easily projected onto non-Western societies, because they are influenced by strong images. The influence of literature, film and other cultural artifacts on studies has been widely described and criticized. Efforts were even made to turn this signifying force productively. Donna Haraway has written on the construction of science, the tropes and figures used, and she has pointed to the possibility of appropriations of such power techniques (1997; 1985).

Regarding surveillance studies, Peter Marks (2005) gives an overview of recent utopian and dystopian films, that have been influential for the production of popular imaginations of surveillance. But it is not enough to admit cultural artifacts are influential. There is a need to understand which tropes influence and structure surveillance studies in which ways, and what effects they produce: a theory/concept of signification, that sheds light on the traffic these transpositions produce in the signifying realm.

Here, only one trope is briefly discussed to provide an example for the problem. When Gary T. Marx introduced the figure "Big Brother" into the relatively new surveillance discourse in 1985, it was already a questionable move to paint such a dystopian picture for the sake of analyzing and understanding surveillance issues. By now one cannot emphasize enough the misleading problems of this transposition.

Once made, this transposition became hegemonic within the scholarly surveillance discourse. This signifier connects seamlessly to all sorts of paranoia and fear. It has become *the* universal trope within surveillance studies. The trope "Big Brother" persistently pushes the imagination to some kind of *totalitarian surveillance* of an all seeing eye, a centralized instances, which knows everything about its targets.

The "Big Brother" trope unifies surveillance, control and the advent of a society void of freedom for no clear analytical reason but its power as a figure in a narrative. As long as it is prevailing and actively or latently structuring this discourse any concept of 'surveillance societies' which is not critically emancipating itself from this burden is already fixed in its trajectory. Additionally, it reproduces an understanding of subjects as static, passive – victims.

Agency, I argue, can be altered, redefined, but never annihilated by surveillance. In a country under heavy surveillance, like South Korea, activists continue their struggles, albeit surveillance is intense.

> The police regularly collect information data to use as evidence, like film. And they can use this to arrest you and persecute you in the court. This is one aspect. Also they are eavesdropping on our cellphones conversations, so we don't decide anything via phones. Also they are watching our internet activities. We have caught signs that they are doing this. But there is basically nothing we can do about it. They are also secretly collecting confidential information and accumulating this against us
>
> Two Korean Activists

From the perspective of an analytics of government, surveillance is a proof of the existence of some sort of freedom. In a state of total domination, surveillance would lack its reason.

The kind of surveillant "seeing" in Orwell's narrative equals total knowledge. Such that the rationality and episteme of this "Big Brother" scheme are beyond a comprehensible dimension: it is an secularized (evil) God. Orwell's "Big Brother" is invested with a super-human mind's capacity to interpret situations and, most importantly, language. Although semantic computational analysis has made significant progress in the last years, it remains categorically a different regime of interpretation, producing false positives and false negatives by its very own systemic limits.

Maybe the word "surveillance" as such already implies such an overlooking figure as it literally means 'super vision', the figure of a sovereign overseeing her territory in totality and detail.

Orwell or Huxley?

A second famous dystopian novel, Aldous Huxley's *Brave New World* (1932), I argue with caution, is much more insightful. While *1984* portrays a repressive regime and its society, *Brave New World* pushes the dystopia somewhere else. The sovereign's subjects are characterized as affirmative subjectivities: from fertilization of the ovules to starvation, 'human' life here is a product of a massive and ubiquitous biopolitical program. Inhabitants of this world of 'happiness and joy', where mourning and critical talk are commonly regarded as unethical, incorporate the values and practices of a regime void of discontent. Thus, the novel is an eye-opener towards subjects and their involvement in governing, in many ways this has analogies with the concept of governmentality and the program of a self-governing society.

Here, the expressions of discontent or dissent are not forbidden, as in *1984*. But they are not very useful either. The key political technology, affirmation, renders them toothless within a distributed, dispersed and dissipated power that is productive within all micro strains of sociability. Huxley offers a picture of a type of subjectivity, illustrated with architecture, leisure zones, industrial production and scientific research, which places the possibility of dystopia away from surveillance-technological developments and the model of the panopticon. Huxley's strength is to propose that without the governed subject's commitment, no regime can prevail. *Brave New World* emphasizes the historical contingency of truth-production and thereby how values and ethics are shapeable, changeable and adoptable. It can be read as an dystopian account of power relations, whose most prominent roots are in each and every person's mind and body, and as such offers a promising starting point for reflections on any past or present regime, its sociality and its surveillance. The statement is that sociality is stable if it can rely on *imma-*

nent factors without expending much energy on repressive measures, just like the concept of the conduct of conducts.

Nevertheless, *Brave New World* never achieved a prominent position in the surveillance studies discourse that can compete with the figure of the "Big Brother". This is a hint towards the limited scope of how surveillance often is conceptualized.

Surveillance and Location

> I have a nephew who is teaching in the national defense college. He is telling me: why are you using phones? It is so easy to track you. He says, they have the equipment and put it in a car and they can get all the cell numbers of the people at some place. He says things like this. He is telling me not to use the cellphone.
>
> Mina, Minerva, Joan, Julie / Manila

Returning to Wood's history of surveillance studies, he asks, "when societies other than the places where Surveillance Studies academics are concentrated are examined, one can see similar developments. But is this as simple as Anglo-American conceptions spreading to other nations?" (Wood 2009, 187).

This skepticism towards global approaches in surveillance studies is of fundamental importance for the research field.[8] While Wood takes the example of Japan (Wood 2007) to show that different societies in different regions develop different behaviors and practices towards surveillance, Minas Samatas (2005) reports on specificities in post-authoritarian Greece:

Identifying four distinct surveillance periods in Greece's post war history, Samatas shows how different surveillance regimes accompanied the Greek people from 1950 until now. Due to the experience of authoritarian surveillance, especially older Greek citizens "are generally unconcerned with forms of direct, physical watching and monitoring of individuals":

> Part of the reason for this is cultural. In Greek language and culture, surveillance (*parakolouthisi*) refers to a natural activity in human affairs. It implies a situation of direct co-presence and the visual observation of people, which is not perceived to be embarrassing or threat-

8 For an analog critique of the notion "global civil society" with reference to news media, see Dencik (2012).

> ening to social relationships. [...] This is in marked contrast [...] to 'filing' (*fakeloma*), the recording and sorting of citizen's information in police files [...]. Direct experience of this type of draconian control system has created heightened sensitivity to processes of 'filing' more generally. This sensitivity to surveillance as 'filing' as opposed to watching continues to the present. (Samatas 2005, 188)

Taken only this one specific report into account that demonstrates a specific, maybe even unique encounter with surveillance, the question arises as to how a global surveillance society can be conceptualized.

In their *Surveillance & Society* editorial text for the issue "surveillance and resistance" Luis A. Fernandez and Laura Huey also argue that surveillance studies need to be situational, contextual, and historically specific, because "[r]ather than categorize the different types of surveillance techniques, then, it is more useful to enumerate the different types of actors who may engage in surveilling a subject" (Fernandez and Huey 2009, 199).

Wood argues that the focus should be "on how particular places are constructed in the encounter with surveillance, and to unpick the scalar elements of this story" as "surveillance operates and operates differently at different socio-spatial levels" (2009, 189). Not only location, but also scale is a parameter in surveillance analysis.

The Concept of Institutionalized Mediation

Foucault's "Discipline and Punish" (1977) is *the* hallmark of reference for surveillance studies. Wood reports that "for Surveillance Studies, Foucault is a foundational thinker and his work on the development of the modern subject, in particular Surveillir et Punir [...] remains a touchstone for this nascent transdisciplinary field" (Wood 2003, 235).

That this focus became a problem in the recent years is not ignored. In a text from 2003, David Lyon agrees that the "older metaphor of 'Big Brother' or the panopticon, redolent of heavy handed social control, seem somehow less relevant to an everyday world of telephone transactions, Internet surfing, street level security, work monitoring, and so on" (Lyon 2003, 18).

To problematize the strong influence of Foucault's genealogical account of the discipline society within surveillance studies, Michalis Lianos has argued that "the Foucauldian model of control, and consequently its explanatory power, refers to the past and is not concerned with the emergence of the contemporary postindustrial subject"(Lianos 2003, 413). Lianos does not refer explicitly to governmentality studies in his critique. But (maybe for that reason) he proposes focusing on the contemporary mediations of cultural and social negotiations in his own way. He proposes that mediation of sociality has seen a shift towards institutional control in a broad sense: from ATMs to web portals to car-parkings, "we live through a multiple range of institutional activities that monitor and verify conformity" (2003, 414). According to Lianos, this institutional control is a "planned managerial activity corresponding to the complex mode of organization of contemporary Western society" (2003, 415). With this in mind, a dispositif like CCTV appears to be complex and multifunctional: it is as much a repressive, disciplining surveillance technology, as it is about detective surveillance, regulation of traffic flow, support for planning, accident-prevention and much more.

More generally, institutional activities establish a regime of fixed and nonnegotiable conditions that mediate and fragment sociality. Lianos most interesting argument is the power to fragment:

> Institutional control is neither intersubjective nor group-based. On the contrary, it is by definition *impersonal* in its origin and *atomised* in its reception, because it is conceived and applied by an institution as part of the homogeneous distribution of a certain activity. Between the institution and the user, there exists no interaction, except for a managerial monitoring and feedback cycle." (2003, 416; emphasis in original)

And as such, he resumes, it focuses "on the organisation and the contextualisation of what is often intended or even desired by a sovereign subject" (2003, 416). Institutional control in this broad sense mediates sociality by directing and setting the conditions of all that it mediates. I understand this concept is wide enough to include money into this scheme, as it mediates sociality and is institutionally controlled from production to distribution.

The individual is determined in her possible social action by these settings, which exclude cultural negotiations and reduces sociality to a binary scheme of *access given/access denied.*[9] According to this concept control is situated in the specific historical setting of managerial activity, which is structured by the processual and resonating character that technology and the social together and reciprocally produce.

In the outcome the subject is desubjectivated and objectivated as her opinions or values are not of concern, it is about managerial problems. This is were this dispositif is productive. It produces desocialisation by imposing non-negotiable normativity, culminating in "the rapid disengagement of users from social belonging" (Lianos 2003, 425).

> We can best comprehend the transitions in control as part of the massive changes in the area of contemporary sociality and the broader atrophy of the social universe. Besides putting an end to the long and imaginative demonisation of control, this angle links the emergence of institutional normativity to overarching social change. It is not control that becomes desocialised, but sociality which institutionalises itself; it is not surveillance which intensifies and spreads, but the demand for systems and networks which propagates itself and favours the fluid – albeit atomised – channeling of individuals. (Lianos 2003, 425)

Control and the social form a new paradigm themselves. They productively engage in a new societal setting. I read "sociality which institutionalises itself" as a deep interweaving of governmental rule and subjectivities. What Lionos calls "institutions" covers all entities and actors that impose a pre-negotiated rule onto subjects in wide areas of sociality. Institutions here are the different programs, political technologies, and institutions in a proper sense, which regulate sociality on a micro level, that *effects and affects* subjectivities.

If sociality institutionalizes itself, it does so as a dynamic that, indeed, follows specific trajectories, but their outcome is not determined as such. There is no need to follow Lianos' view in its entirety and it is disputable whether this dynamic should be called "atrophy of the social universe".

Nevertheless, drawing on the institutionalization of sociality which in effect favors atomized individuals over mediation of groups helps considerably in understanding that surveillance itself cannot be understood as an isolated practice, but

9 See also Deleuze (1992) .

that it is embedded in a more complex web of mediations which structure sociality. This structurally enmeshed aspect is of importance and often overlooked. Lianos has shown this.

From here, two concepts of contemporary surveillance are introduced, which are useful for a discussion of data retention.

Panspectron and Surveillant Assemblage

The concept of the *panspectron*, as introduced by Sandra Braman (2006b), resonates well with Lianos' concept of institutionalized mediation. It refers to a state of things where no surveillance subject is specifically invoked in order to trigger an information collection process. Rather, information is collected about everything and everyone all the time. An individual subject appears only when a particular question needs to be answered, triggering data mining for particular information within the mass already gathered, in order to precisely answer that question. And while populations remain generally aware of the unmoved and intimidating presence of the panopticon, they tend to be unaware of the aggressive efficiency of specific modes of information collection. Data retention exactly fits into this conceptual frame, just as Passenger Name Records (PNR) and SWIFT financial data. These are sustained and augmented without any specified trigger, and therefore, potentially infinite.

A different, but again in large parts compatible approach towards surveillance is the concept of the *surveillant assemblage*, as introduced by Kevin Haggerty and Richard Ericson (2000). It refers to a multiplicity of heterogeneous objects, whose unity is solely functional. As an assemblage is always a "potentiality," this concept can be connected to the developing panspectron, which also resides in the background as a formidable and ambivalent latency.

Paradoxically, the diversity of the assemblage is also its power: "As it is multiple, unstable, and lacks discernible boundaries or responsible governmental departments, the surveillant assemblage cannot be dismantled by prohibiting a particularly unpalatable technology" (Haggerty and Ericson 2000, 609).

The main direction of the assemblage is to transform the body into virtual bytes of information — data doubles. "And while such doubles ostensibly refer back to particular individuals, they transcend a purely representational idiom." (Haggerty and Ericson 2000, 614) I will return to this very important notion about the transcendence of representation, as this is achieved by the production of a new knowledge-power axis. Discrimination and social sorting is amongst the socio-political consequences that subjects may experience as concrete back-references. Data retention is an example of a functional element of the assemblage that might refer back to subjects. It is unpredictable, and possibly holds radical consequences for the subject. But its internal logic points beyond representation and operates within a realm of algorithmic abstraction.

The following passage is intended as an inlay to the discussion so far. It recounts a larger history of engagement with surveillance problems in an activist and artist scenario.[10]

Interlude: Ricardo Dominguez on Transparent Surveillance

Ricardo Dominguez, a professor at University of California San Diego and long time activist/artist with digital technologies, recalled many of his experiences and concepts when I interviewed him and asked about surveillance. In his detailed and original answer, he touches many of the core aspects. Interestingly, paranoia and fear is prominently placed in the center of his answer. Dominguez stands out from all other people interviewed in the sense that on the one hand he situates himself within the fields of art as well as in the field of activism, and on the other hand he is the only person I interviewed who engages in practice *and* theory at the same time.

His current appointment at UC San Diego led to the development of 'border disturbance technologies' such as the *Transborder Immigrant Tool*.[11]

In an unconventional manner, I have decided to reproduce his narrative about surveillance in full length, as it enfolds along a narrative of practice with different groups and actions and develops a complex argument which I hesitate to shorten.[12]

---

10 This interlude obviously is not part of the methodological framing laid out earlier on.

11 See http://bang.calit2.net/tag/transborder-immigrant-tool/.

12 As Dominguez refers to a couple of projects, groups and special terms in his discussion of surveillance, I have added them to the glossary. Nonetheless for a better reading they are listed here, too:

I began to think about surveillance in terms of its uses probably somewhere in the mid 90s. In the 80s you would do actions and the cops would be there watching you. So, that was kind of self-evident surveillance. But I think there was a theoretical fiction by William Bogart, that came out sometime in the 90s, that is called "Simulations of surveillance". I found that text be quite useful in thinking this question, about mass surveillance, distributed surveillance, the histories to the panopticon. But again, the question was in the same way that Baudrillard says "surveillance is bad", Virilio says "speed is a negative condition for our society". And Bogart certainly focuses on bringing surveillance to this kind of negative space of society. One of the core elements in his story is: what is simulation of surveillance? The story he tells is: Out West, say in the spaces in the desert where I am from, there is a small town that can not effort having police force slowing you down. So your are driving for hours really fast. And you make a turn and suddenly there is a cop car. So you have to slow down, watch yourself, and all that. But you may not look and it is really a dead car, painted to look like a cop car. Because the town can not afford having cops sitting there all day long. There you create this kind of simulation that you are being surveilled on a No-Fi level. It is that sort of agglutination of aggregation of a panopticon. And you no longer have to create the prison, you only have to create the dead object that evokes that condition of surveillance. And the reason this question became important, at least to me, was the emergence of the practice of Electronic Civil Disobedience. Since the choice that we had to make when we initiated the direct practice, rather than the conceptualization, was that should we seek that kind of history of hacker and anonymity under a name. And certainly collective names have been important in my history of CAE and EDT. But I felt that we needed to do was to kind of disturb this simulation of surveillance and that was to not be anonymous. To be wholly transparent. Because this would fit in a clear way with the Civil Disobedience model, that we were trying to teleport. It is the body, in the street, that signifies the authentic condition. So how could the cops give us a beating if they didn't know who we were? It was a bit of internal struggle. A good part of the group came from

---

- Critical Art Ensemble (CAE)
- Department of Defense (DoD)
- Distributed denial of Service (DDoS)
- Electronic Civil Disobedience (ECD)
- Electronic Disturbance Theater (EDT)

the tradition of the freedom of anonymity, of having the technology itself speak rather than the person. But I felt that we needed to say: I am Ricardo Dominguez, I live in Williamsburg, here is my telephone number, if you have any problems with us coming to your website next week where we do a virtual sit-in, please contact us. I thought this would disturb this condition. And create a counter simulation of surveillance on them. And that they would have to participate in announcing "Hi, we are here". We are the FBI or the Department of Defense. So, they themselves could not specifically hide, because it was a transparent game. And of course the question of transparency is also very much kind of a tweak of the neoliberals to cause all kinds of problematics. At that particular juncture I thought it was important to seek out surveillance, to interpellate me: I am over here officer, digitally. And I am not suggesting that it is the best thing to do with life outside. But digitally, being targeted for surveillance, because again it was a way. At least that was my gut feeling: not to be concerned with surveillance. Because I was giving them all my points, my telephone number etc. So, when the Mexican Government where calling when we were at Ars Electronica doing "Swarm" in 98: "We know were you are, we know what you are doing." I said: "Of course you do! I just posted it all online. It doesn't take a deep intelligence system to figure it out." It was more transparency: to evoke surveillance was to evoke those powers to have to manifest themselves. Whereas if one is in some kind of paranoia, whether it is simulation of surveillance or surveillance proper, through this kind of transparency one is not concerned. In a certain sense one can see the activists of the 70's have been driven insane: they are watching us! Paranoia: you are not sure. And it seemed to me that they are being paranoid about being watched. So I said: if you are that bored: go, read my email! That was kind of trying to disturb that kind of surveillance model and create a minor simulation of transparency, which would disturb the major simulation of surveillance. The major components of Echelon. We used to do virtual sit-ins against Echelon. So, this matrix, then traces out, takes measure of this kind of social models of surveillance. And to do it without any kind of fear, which I think is often the way how surveillance is used. And so, that has been the model. For instance with the Transborder Immigrant Tool, you get email all the time "you are a traitor, you should be beaten". But then once again: one allows a public sphere. They might be anonymous, "White guy number 89 says". But then this brings out the different players in a way that is not a hidden component of the fear or secrecy of surveillance. Surveillance is part of the space of public fear. To create a disturbance in the discoursive formation of surveillance, in the kind of technological effectiveness of surveillance and cover and take measure of its different components. I suppose the argument could be made now that we have a move from transparency in toto to one of translucency which is to have a kind of movement between transparency and opacity. Eduard De Glissant, the Caribbean theorist, speaks about the politics of the opaque. Especially around communities of resistance in the Third World. Certainly, if you are a young Filipino fighting police abuse and you are doing Electronic Civil Disobedience, I would recommend opaqueness. So, I am taking it into account that I am a certain citizen of empire, established within a certain circuit of civil rights, limited, but certainly elements, so, I have a matrix of performativity which is much larger than other communities. It is a privilege. So, when I speak to communities I do take this into account. This performative matrix of transparency in relation to surveillance and opaque politics have their space. In a certain sense there is also to me this third paradigm of translucency, which is: if there is major simulation and there is minor simulation in Deleuze' notion of "What is minor literature?". And so, I feel that the work that ETD does is one of minor simulations that riffs of between transparency and translucency. Does this work? Is CED really working as practice? Well, the Mexican government seems to feel it is working. So it creates this possibility. Usually the feedback is quite critical. When we did the action of "swarm" at Ars Electronica, nettime, with all these powerful and important people, deep thinkers that we had learned a lot from, said "no! activism is about communication and documentation." So this direct transparent action is kind of destroying a lot of the work that has been done. And it is a paradigm that is not useful, is what they said. Activists on the ground. There are cyber borders and you are a Northern group moving into the South, once more, this kind of post-colonial attitude. But on the other side, the Zapatistas called for an intercontinental, intergalactic campaign. When they came and spoke to us, they said it is fine. So, are there really these sort of conditions? The hackers, like "hippies from hell", at that time they were heard and they said: "this is distributed denial of service, it is not digitally correct." And we said: "OK, so we are digitally incorrect in our process." "It is a useless technology" they would say, "ineffective in the process." But for us this uselessness and ineffectiveness was an affective politically coordinated element. So, there was a lot of intercourse, the institutions of power, the DoD, all of that, were also extremely angry and they were saying "no". So we were able to trace out the conditions of power, minor powers, majors powers all around this question, and those critiques were very useful in implementing a larger scale performative matrix, that disallowed the end of the practice, but also disturbed each of the discourses and allowed perhaps a minor conceptual drift, that was unexpected for each of them, which then let the Congress saying "ECD is not cyber terrorism, is not cyber crime, is not anonymous

> cracking, it is a teleportation of Civil Disobedience. Nothing less, nothing more." It is part of the practice of performative matrix of ECD at that point which is to look at the kind of emerging hard definitions of cyber war, cyber terrorism, cyber crime, and shifted from questions of infrastructure and syntactical knowledge code to semantics. And semantics was a way to reshift and teach a discourse. Wherein communities of power and minor communities of power had to deal with the semantic structure as the primary shifter point, rather than the question of syntactical code and infrastructure. Those were the things they said: it is stupid to DDoS, it is illegal. And all of that. But that is not a landscape one can negotiate pedagogically. And shift that power. So, with minor simulation, transparency of surveillance one could then have that pedagogical shift and certainly we saw it starting in Germany with the Deportation Class Action and the legal outcome of that case. The law case said: they went with the cyberterrorism/cyber crime model, the higher court in Frankfurt said "No. It is one of public protests." So, I think that was an important discoursive shift.
>
> Dominguez / San Diego

One important point for the discussion here is to demand that surveillance, which is mostly an opaque operation beyond feasibility, becomes visible, to demand it to come to light with the means of a "minor simulation of transparency". This is an answer to the problem of paranoia and the silence it produces. It can be tried, as Dominguez indicates, when the social position allows to do so. To engage surveillance into a conversation can be achieved by demonstratively being transparent in actions that provoke special attention to authorities. Doing this in the digital realm converts the opaqueness of computer mediated actions into non-digital phenomena: the call or the email that responds to the irritation caused is a demonstration that surveillance exists and has to show itself. Simulations of surveillance cannot react on such a call. The production of a vis-à-vis situation with surveillance turns the general idea of surveillance upside down. The discursive shift changes coordinates and suddenly new practices become part of the discussion, like virtual sit-ins, which so far had been treat as hacking and terrorism. Now they are legalized and protected under the right to assembly.

Nonetheless, in the next chapter I descibe data retention as a political technology of sovereignty that answers to the liberlalized dispositif set up by mobile media. Additionally, other occurrences that are part of the large conglomerate of mobile media are introduced. With this, I try to provide a different approach towards mass surveillance, one which is situated within an analytics of rule and describes different political programs as intertwined although they follow heterogeneous trajectories.

7 Re-Entry of Mobile Media Sovereignty

> We pay a lot of attention on surveillance, there are many situations where people have to turn phones off and take out the battery. In both, mobile and non-mobile, surveillance is a big issue.
>
> MST Workers / São Paulo

I have argued that the massive dissemination of mobile media affordable for large parts of the population contributes to governmental regimes of power. Governmentality, as I have laid out earlier on, coincides with the production of 'freedom', whose means are the apparatuses of security. Foucault is concise in that the

> mechanism of security does not function on the axis of the sovereign-subjects relationship, ensuring the total and as it were passive obedience of individuals to their sovereign. They are connected to what the physiocrats called natural processes, and which we could also call elements of reality. (Foucault 2007, 65-66)

Apparatuses of security are productive, they "work, fabricate, organize, and plan a milieu [...] in which circulation is carried out" (Foucault 2007, 21). Mobile media technologies, its devices, its infrastructure, its economy are all results of the organization of such a milieu, where, plain to see, circulation of communication is carried out.

While such an understanding of (tele-)communication – as securing freedom – is intuitive, the techniques, measures and technologies that reduce, limit, and complicate circulations of speech and signs mark a different point of interrogation, all the more when they appear *after* the population's massive adaption and use of it.

To limit mobile media through governing marks a tipping point of the inital freedom of mobile media. It shows that the common reference to mobile media as an empowering technology, as an facilitator of agency, needs to be framed within a play of modalities of rule. Finally, this gives an occasion to understand mobile media as surveillance media whose trajectory is to control the freedom mobile media previously had produced, eventually turning into an illiberal technology.

Examples of such instances are data retention schemes, mandatory registration of SIM cards and blocking/jamming of cells of a mobile network. What these cases all have in common is that the free circulation of signs is confronted with a political technology of control that in different ways disrupts the flow or reformulates the flow of signs.

All of them, in varying degrees, affect large groups, up to the whole population. None of them is targeting specific individuals, although data retention creates the possibility for this, as well as SIM card registration creates the possibility to identify targets. Additionally each of these technologies has emerged after the massive roll-out and of mobile technologies and after it has affected the population. Therefore, in a sense, they all answer to the production of freedom by an apparatus of security; to an expression of liberal rule. They mark the paradox of liberal rule, which has to intervene in the program of a "self-regulating society", although it shall not, at least according to neoliberal ideology.

These measures overrule the liberal frame and establish a preventionist scheme (data retention, and in a sense as well SIM card registration), they limit it as they discriminate between those who are eligible and those who are non-eligible (SIM card registration) or locally shut down the circulation of signs in its entirety (blocking of cells). Each of the cases performs on a different field. The discrimination they perform on is in the temporal (data retention), social (SIM card registration), and the spatial (cell suspension).

Clearly, such instances share characteristics of sovereign power: the state of emergency enacted by the Musharraf regime in Pakistan falls within sovereign rule, as it revokes the constitution and thus liberal rights granted to citizens are annulled. Media is banned, public gatherings are forbidden. Sovereign acts abandon governmentality programs, regimes of government that have rationalized economy: they contradict the paradigm of self-regulating societies. But as well, there are specifics in each of these cases that hint towards something else but 'classic' sovereign rule, as they operate within the informational milieu (Terranova 2004).

By discussing each of these examples separately, I will try to identify, within a framework of an analytics of government, what mode of rule they initiate and how this relates to the problem of securing freedom, to the production of freedom, and to channeled actions which are the ground for the conduct of conducts. To my understanding, all of them are different problematizations of a flow of signs, goods, and bodies. With different means, they all work on the opposite end: restricting and controlling a circulation that liberal regimes had initialized. As "security [...] tries to work within reality, by getting the components of reality to work in relation to each other, thank to and through a series of analyses and specific arrangements"(Foucault 2007, 47), one may ask in what way any of these three cases supplements security. All of them intervene into the liberalized mobile media scheme. But before the discussion of the three cases, I want to reconsider the term "sovereign rule" and why this is important.

Illiberal Government and the Surveillant Assemblage

One might simply say: 'There is all this liberalized technologies that float around, everyone is using it. So, evidently, also the bad guys use it. That's why state has to intervene and secure the apparatuses of security against a wrong use.' Although on the surface, such a claim might not be wholly wrong, it lacks analysis in many ways: first of all: who are the bad guys? And what is this state that intervenes? There is no such thing as a sovereign that overlooks her territory. The state, and this is one central claim of the Foucauldian framework, is not a homogeneous entity. Addressing it at such does not help to understand the myriad of semiotic-material conglomerates of rule, which allow the performativity of government.

Sovereignty usually becomes subsumed under the term 'the law' or the execution of law. Foucault left open a thorough analysis of sovereignty in his late work in regard to neo-liberal societies. He mentions the prevalence of three pillars of power in contemporary rule, but by focusing on (neo-)liberal rule, one could get the impression that rule is all about the production of (economic) freedom by apparatuses of security and the production of the economic subject.[1]

1 In his lecture from April 4th 1979, Foucault returns to the problem of sovereignty vis-à-vis the *homo economicus*. He explains that the subject of rights and the homo economicus are irreducible, because the sovereign can only fail to oversee the totality of the economic field. Therefore "the emergence of the notion of homo economicus represents a sort of political challenge to the tra-

This neglected point is what Sven Opitz refocuses on: the role of sovereignty in governmental societies. He looks at the "re-inscription of sovereign logics of power into governmentality as liberal practice" (Opitz 2010, 94). His focus is on a new mode of sovereignty: *securitization* that "calls for direct interventions which ignore the bounds of law and install a sovereign exception to liberal rule. Under conditions of modern governmentality, the concept targets the point at which liberal regulation—in the name of liberality—becomes illiberal." (Opitz 2010, 101).[2] What Opitz discusses therefore is a form of *neo-sovereignty*, one that operates within governmental rule, but one "that renders the illiberal moment of the security dispositif dominant, and thus functions as the decisive hinge that allows the intersection of liberal and sovereign modes of power" (Opitz 2010, 101).

There is an important clarification one has to make here: because societies of governmentality have become an object of study in broad and many ways, one has to be aware that the production of self-governing subjectivities as its base *only* follows the trajectory of outsourcing the risks of economic well-being to the individual, leading to the redemption of concepts of solidarity for societal relations. (Neo-)liberalism is for and most an economic program that has rendered economic thinking the paradigm to govern. It is important not to fall into the trap of this episteme oneself. Disciplinary powers, as well as sovereignty in the form of executing the law, have always sidelined governmental rule. Where subjects of self-conduct failed, disciplinary measures and the execution of law are necessary to avoid a void: the "enterprise society"(Foucault 2008, 147) must comprise the whole population. What is more, the imperative of the market decisively attacks all that is external and through its trajectory of "centrifugality" (Foucault 2007, 45) increases its field of operation. Therefore the problem that governmental rule faces throughout is its own economic enterprise: it is haunted to finding the right balance as it shall not intervene into the pursuit of its governed and it must not interrupt

---

ditional, juridical conception, whether absolutist or not, of the sovereign" (292-293). It is *civil society* that appears as the new frame of reference, or "plane of reference" on which government has to be performed. Thus, Foucault presents the "transactional reality" (297) through which liberal government integrates the subject of rights and the subject of economics, but leaves open how this new plane of reference relates to sovereignty (Foucault 2008, 291-316). For a discussion of a shift within Foucault's work between "Society Must Be Defended" and the lectures of 1978 ("Security, Territory, Population"), see Collier (2009).

2 The concept "securitization" was developed by the so called *Copenhagen School*. See Buzan, Wæver, and de Wilde (1998) and C.A.S.E. Collective (2006).

the flow, because this causes high expenses, and if it intervenes too late, or at the wrong places, or with the wrong programs, expenses are even higher.

Thus, essential for government is the right balance of the three forms of power, that liberal rule is always in the course of adjusting.

The "play of transactional freedom" in itself is immanently incapable of integrating all individuals. The idea of a self-regulating society that economical rationalities prescribe, obviously has systemic deficits. First and foremost as it is a totalization of economy over the idea of society. This is were the paradox of liberal rule unfolds and securitization and other means of illiberality 'blossom'.

This means that one can speak of political technologies of neo-sovereignty, which are not entirely or only in an indirect way connected to lawful action, and which are pursued both within the governmental neo-liberal program, and in contradiction to it, as such interventions are faced with the strategic problem not to halt circulation. They are materialized expressions of the liberal paradox itself. "Government must not form a counterpoint [...] between society and economic processes" (Foucault 2008, 145). As society is conceived as within a regulatory principle by the mechanisms of competition. Such an "enterprise society" is in need of its own kind of sovereignty: its programs can be called securitization. They operate outside of lawful practice or at least in an (yet) undefined territory; a gray area, where police, the army and the recent omnipresent 'security services' lay claim on the execution of a sovereignty, which, and this is important, operates either in addition or beyond juridical sovereignty.

Episteme and Modalities of Neo-Sovereignty

This neo-sovereignty sets up a boundary: that between the "level of the interplay of differential normalities" (Foucault 2007, 63), which defines the realm of governmental rule, and what exceeds this differentiation to the degree that it cannot be "normalized and needs to be confronted illiberally. Thus the erection of a barrier, beyond which government can grant no latitude for freedom, correlates with a discrimination of possible and impossible subject positions"(Opitz 2010, 104). The anormal is a correlation of differentiated boundaries; there is no static anormal that triggers liberal or illiberal rule. As the technological measures on both sides operate on statistical sizes I propose to understand both sides of the barrier as a dynamical space and the boundary itself as moving. This is an expression of a logic of risk, which cannot refer to an 'ontological' or 'static' episteme, but is always in flux and addresses no concrete individuals, but statistical normalizations. Setting norms thus can only come under way within a dynamic set of parameters. Lawful action is not part of the parameters itself, but rather subjected to the statistical normality produced.

More to the point, I suggest that there is no, or not always, a "hinge" as such, no clear switching point, but a rather field of statistical approximations which lead to the invocation of illiberal rule. This episteme that produces the truth for government is largely a calculated evaluation via algorithmicity and big data; an error prone risk assessment reified as computational objectivity. For example, when analyzing traffic data of communications 'suspicious' patterns of association can be produced (Strandburg 2007). While these patterns as such can be a result of many things, its simple emergence already causes suspicion. When such weak ties become the operational base for government, as it is objectified through algorithmicity, the effects of governing can face severe contingencies.

Eli Pariser calls the shaping of the future in the context of the 'filter bubble' that the internet, he is arguing, is about to become, "algorithmic induction", which "can lead to a kind of information determinism, in which our past clickstreams entirely decide our future" (Pariser 2011, 135). The problem he addresses is called "overfitting" and leads to "a regression to the social norm" Pariser (2011, 128). In parallel, I suggest, this epistemic logic is inherent to algorithmicity in neo-sovereign programs.

Seen from a broader scope, data retention fits well into the post-9/11 *war on terror* instruments. The shift toward an omniscient surveillance-state has generally often been compared to scenarios familiar from the prophetic novel *1984* by George Orwell (1949). But there are critical differences (see chapter 6). The analogy had been underscored by the historical concept of the panopticon, introduced by Bentham (1995) and popularized by Foucault (1977). But where the panopticon draws its power from the fact that the surveilled never know if they are surveilled, and therefore internalize habits *as if* they were surveilled, the present situation, fostered by ongoing modalities such as data retention, should more accurately be referred to as *panspectron* or from a different aspect as *surveillant assemblage*, because everyone who is telecommunicating is always under surveillance, although, and this is what I want to discuss further, this surveillance has its very own characteristics.

> So, the thing that around mobile phones the most shocked me was when it was revealed that the US carriers where surveilling citizens without warrants, AT&T and these things. That was a big knock. Not that I was that surprised, but that was just so common. That really led me to thinking that surveillance is more of a day to day problem.
>
> Freitas / NYC

It has strong and specifc effects how, or even if, the 'anormal' is produced and identified to a large degree, as the 'threat' is constructed in a relation between two differentiated statistical approximations: the 'positive' liberal one, and the 'negative' illiberal one. Thus, contingent decisions have to be made between these two in part overlapping data sets. These contingent decisions are part of the algorithmic regimes, the machinic operations at work. Matthew Fuller has termed the bits and pieces about an individual in a database[3] that produce new data objects *flecks of identity*:

> [I]t becomes possible to describe *flecks of identity*, in their existence as standard objects, within databases as a primary compositional element within surveillance systems. This is what at its scalar levels control *sees*, an informational token of conformity or infraction. An element, cluster, or concatenation of data, flecks of identity – a number, a sample, a document, racial categorization – are features that identify the bearer as belonging to particular scalar positions and relations. [...] The citizen has a place, a speed, a set of functions as a variable within a social, bodily, technical algorithm. (Fuller 2007, 148; emphasis in original)

Which flecks of identity are queried and how they are combined is perspectival and depends on the security query done in the database and the algorithmic operations triggered. In the end, the subject is flagged or not flagged. This binary event stands in a drastic disproportion to the data processed. The operations that lead to this state are infinite, centrifugal and opaque. "Gender, race and sexuality have

3 About the database as a key form of the computer age that correlates the narrative as a key form of the novel and the cinema in modern age, see Manovich (2001).

been reduced to recombinable elements, disassociated from their subjects and re-composed on a plane of modulation."(Terranova 2004, 35) Neo-sovereignty is neo because it operates on a plane of modulation, which lacks the nexus to the material that provided such elements in the first place to non-algorothmic sovereign operations.

> A good fellow of me, whose father is a policemen, once told me that I was filed and it was very detailed. Very many people are filed, and the police likes to collect information about people, even people who have committed no crime. They are always just adding. The files are getting bigger and bigger, more and more detailed. And they collect from both: analog and digital.
>
> Fernão / São Paulo

Flagging Data Flecks

An example for such a 'flag' operation are 'trouble maker databases' that have been produced in the recent years in different member states of the European Union. The aim is 'prevention and protection against serious threats for public security', as the delegations that discuss such matters continue to say.

For Germany, in 2009 a database has been set up called "violent sport offenders" (*Gewalttäter Sport*).[4]

> Although the file's name suggests that it holds information on violent offenders, many of its entries do not refer to individuals who have been convicted of a crime but rather to people who have received a ban or were subject to stop and search procedures at football matches. (Töpfer 2009)

---

4 Within the European Union databases on specific subgroups of the population are numerous. Some are interoperable, some are not. Some are accessible by national police, some by federal police, some by secret services. Some store the data unlimited, some have official deleting schedules. I will not discuss this labyrinth of control and surveillance in toto, but only refer to the "troublemakers" databases, as it targets specifically football fans, who by nature come in masses, and political demonstrators under the de-politicizing label "troublemakers". For the general trajectory, I agree that "[s]ince the onset of the EU's response to the 'war on terror' the prime targets have been Muslim and migrants communities together with refugees and asylum-seekers. Now there is an emerging picture across the EU that demonstrations and the democratic right to protest is among the next to be targeted to enforce 'internal security' " (Bunyan 2009).

The realization is opaque and contingent: the risk assessments done by individual police officers, who decide for the storage of personal data in the database, which then is translated into objectivity by the very machinic operations. Once fed into the database, one is nonnegotiably flagged. Illiberality blurs and blackens out, where due law once was the standard procedure. Governing is becoming opaque and black-boxed for the governed.

> Anyone whose data is stored in these databases might experience serious consequences: their freedom of movement might be curbed when they are ordered to register in-person at their local police station on a daily basis (e.g. for the duration of international football competitions), when they are prohibited from leaving the country or when they are visited by police at so-called "troublemaker addresses", in their homes or at work. Moreover, their patterns of movement might be profiled and discreetly recorded at police checkpoints. (Töpfer 2009)

On IGAST, a BKA operated database on 'violent troublemakers who are active internationally' it is known that in June 2009 "information on 2,966 persons was stored in this database. Only ten per cent of the entries refer to 'potential troublemakers', (i.e. those who have been arrested or registered in the context of violent protests against economic globalization in Germany or abroad). All other entries are on contacts, witnesses or police informers."(Töpfer 2009)

In terms of deviance these flagged subjects are only in part illiberalized, they are not expelled or excluded from governmental rule in toto: the calculus of governmental neo-liberal rule only divides the individual, produces a dividuum as Deleuze (1992) puts it. The digital doppelgänger is divided from the undivided and undergoes a process of screening, where the 'anormal' is generated statistically as a 'contingently proven' deviant subject position, based on data flecks without assessing the subject as a whole. In the logic of computational policing, the target of flagging data flecks is a illiberalized but still governmentalized subject. It aims only at a specific reductions of circulation, here the right of free movement or recurring stop and search procedures; the production of deviance is under constant reconfiguration and a highly abstract matter. In this episteme, individuals are not classified according to psychological examination or for motivations which address the concrete person. Instead, flecks of identity provide the means for a construction of a risk assessment whatever the concrete and situated context has been (see also Castel 1991). Deviance is conceptualized as statistical and algorithmical accessible or even producible.

The Limit Figure and Relationality

The primary and traditional figure that organizes the boundaries of the 'powers of freedom', that exerts the 'petty' states of exception, this has been pointed out by Foucault already, is the police: "police is the permanent coup d'État [. . .] that is exercised and functions in the name of and in terms of the principles of its own rationality, without having to mold or model itself on the otherwise given rules of justice" (Foucault 2007, 339). Amongst others, it "acts as a technology of the future, concentrating on practices of prevention that lie beyond the codified criminal laws regulating the prosecution of offenders" (Opitz 2010, 107). Preventionism is an operation based on statistical knowledge for the identification of future crimes and 'dangerous behaviors'. The 'troublemaker database' is a preventionist technology in that sense. By differentiating between acceptable and unacceptable normalities in the statistical sense, technologies of prevention are invoked to identify subgroups of the population as causes to act upon only potentially occurring events. The problem of the factual arrival of an event is therefore epistemically overwritten by its prediction and its algorithmically evaluated probability.

Preventionism invested with algorithmic operations on data sets replace the factual *and* the unwritten future events by claiming knowledge about the future. They prescribe the future, produce actions in the present based on this prescription and thus intervene into a present without a factual reason. This way they alter possible events in the future. The pursuit of life is structured along pre-emptive measures, generated by calculations on the doppelgänger, as "[t]races rather than traditions are what connects body with places" (Lyon 2001, 19).

Ricardo Dominguez emphasizes the powers inherent to relationality as the trajectory of commodity surveillance.

> The data body has now become not just important in its singularity, but in its relationality. I think that dataveillance is part of evaluation and acknowledgment that the individual is only a value in relation to a larger database. It is a database of relationality to the commodity, to the informatic. That is the zone of what tweets you look at, what tweets you don't. And also this question of friendship. This is a loose term for social relationality. Those then become the markers of this new 2.0 economy, which will then either be one of filtration

> through kind of internal commodity surveillance, the commodity itself surveils you.
>
> Dominguez / San Diego

And in the same sense, but more in relation to activism and surveillance, Devin Theriot-Orr states:

> I think the government is actually fairly smart and they can identify services people are using. And there are lots of moles and infiltrators and people who are dressing as activists who are not really activists. So, if I were in the government I would say: we need to be on that. But in terms of threat models to us: relational dataveillance is the biggest.
>
> Theriot-Orr / Seattle

This points toward the discussion of data derivates that Louise Amoore proposed on the e-border systems that are being deployed in the UK: "a risk based system that deploys processes of data mining and analytics in order to derive a risk score or flag for individuals entering or exiting the UK" (2011, 25). At the core of such system governs a calculus of relationality that has data derivatives as its objects. The point is that derivatives are a specific abstraction, "that is based precisely on an absence, on what is not known, on the very basis of uncertainty" such that

> the processes of data integration, mining and analytics draw into association an amalgam of disaggregated data, inferring across the gaps to derive a lively and alert new form of data derivative – a flag, map or score that will go on to live and act in the world. (Amoore 2011, 27)

In a parallelism to financial derivates, these data derivates are forms of correlation which are not centered "on who we are, nor even on what our data says about us, but on what can be imagined and inferred about who we might be – on our very proclivities and potentialities" (Amoore 2011, 28). Data derivatives thus make data actionable as projections. Just "[l]ike filmic projection, the gaps between underlying data items are precisely what makes projected future of the data derivative possible." (Amoore 2011, 29)

This changes how decisions are made. In the case of the e-border program:

> Where the association rules of a piece of software code infer 'who's coming into the country' and 'why they're coming', they release into

> the world a data derivative that intervenes retroactively in order to have already decided the event. (Amoore 2011, 30)

Data derivatives thus are an abstraction from database entries, which operate on a pre-emptive calculus of risk that use the concrete *flecks* for assignments of the future. Thc data cntries themselves serve as "a set of relations from which the derivative can be written"(Amoore 2011, 37). By this, "specific temporalities and norms of algorithmic techniques [...] rule out, render invisible, other potential futures"(Amoore 2011, 38).

In a note sent out to the Police Cooperation Working Party of the EU on 27 February 2009 by the Council Presidency, an interesting proposal of such neo-sovereign measures flashes up:

> The German delegation then made an extraordinary proposal that Member States should be allowed to 'flag' alerts on violent troublemakers even if: such alerts were incompatible with national law, because when there was a 'hit' the action taken would be in the requesting Member State – not in the Member State which lodged the data. This is saying that in a Member State where there is no legal concept of a 'violent troublemaker' the police collect information and intelligence on 'violent troublemakers' and put this onto the SIS then allow the two Member States who have laws on 'violent troublemakers' to access it and use it to take action (possible coercive) against the individual(s). (Bunyan 2009)[5]

Although the German proposal was rejected, it shows the mentality vis-à-vis lawful action in the control and surveillance discourse: here the idea is to stretch the operational field of a database into territories, where this database is illegal. As the place of action would be within the legal territory of this database, the executive part as such would be lawful.

Sovereign power as the juridical and executive arm of the state, just as governmentality, is under constant reconfiguration and pressure, because, like any political technology as a semiotic-material conglomeration, it answers to effects of governmental rule and problematizes it.

5 The two only member states that currently have such laws are Germany and Denmark.

> "[P]roblematizing the *security* of self-regulating spheres marks the tipping-point that makes it possible to navigate the paradoxical relation of non-intervention and intervention within liberal rule." (Opitz 2010, 98)

For the condition of the present sovereignty, one can observe the effects of these operation where subjectivities that are predicted to be not governmentizable can be excluded, expelled, tortured, denied the rule of law, denied the principle of *habeas corpus* and many other phenomena that have emerged in the recent years, often in the rhetorical combination with the 'war on terror'. The other modality of this neo-sovereignty is to define spheres of practice, which work only on parts of the subject's freedom, such as limitation of free movement on specific days.

The screening of the population via large databases is a technology of securitization that operates in the logic of governmentality, as it takes the population as its statistical base for knowledge production. By way of a change of modality though, here illiberality is the outcome: the application of apparatuses of security, as programs for the production of a milieu of circulation, can be rendered instrumental for illiberality. Neo-sovereignty employs these technologies for the production of illiberality, to shutdown circulation for those identified beyond the threshold. This at the same time is an immanence and an excess of liberal governing: the articulation of control, which supplements the apparatuses of security. Therefore the 'neo' or new modality of sovereignty: it is molded into the liberal paradox itself.

After this consideration of the actualization of sovereignty in neo-liberal societies, what it targets (the possible threat) and by which means (production of the doppelgänger), what it answers to (governmental freedom) and what its regime of truth is (the database and algorithmicity), I will now return to the three before mentioned interventions into mobile media's circulation of signs: the socially discriminated (registration of SIM cards), the spatial (blocking of cells), and the temporal (data retention). The data retention discussion will be by far the longest of the three, as it relates in many ways to the larger subject of contemporary surveillance.

Registration of SIM Cards

> They are putting a name and face to every mobile phone. From the government point of view to prevent frauds, mainly blackmailing.

> But in fact regarding mobile phones it is about political control so they can know where someone is, like with GPS.
>
> Alpha / Mexico City

The mandatory registration of SIM cards for pre-paid phones after 9/11 is in many countries generally invoked as an anti-terror 'tool'.

> This law intends to get information about every person that uses a phone. When you buy a phone or make a contract, you have to give your adress, what you need it for, give documents about your adress, birth certificate. A lot of information that you need to give to the private companies. These companies need to save this informations. And they can save all conversations from mobile phones. So if the police wants information, the companies have to pass it to them. This law has passed in February this year. In necessary cases, the companies can record phone calls. This is the national registry of cellular phone users. The government says they do this to protect the population, but it is about control
>
> Olinca / Mexico City

This registration scheme, in the first instance, is a means of knowledge production; not as a statistical tool (although applications are possible) on the population, but rather a classic index, a scheme to produce a legitimacy of mobile phone subscriptions. Whereas prior to such schemes, anyone at any age could obtain any number of SIM cards, the administrative scheme, for the first time, introduces the possibilities to rule on the legitimacy.

But already the invocation of such a scheme excludes parts of the population and discriminates. It is a means of social sorting, because documents have to be provided. The necessity to provide accepted, valid and legal documents excludes everyone without proper documents in the first place, but as well poorer parts of the population, like street dwellers.

> It was easy to get a mobile chip, but now you have to register. This is also for opening bank accounts and landlines, where you have to produce all kinds of documents. So, you have to produce different documents for different purposes. The permanent account number for tax payer is okay for opening a bank account, but not for a phone registration. It is also a measure against marginalized people, as the mobile phone acted as an address, he had a number and could place stickers, but now it has become difficult as you have to produce a proof of residence. This is pre-Mumbay terror attacks, it is from the Delhi parliament attack in 2007
>
> Kumar / Delhi

Connecting the access to mobile phones to ordered residency is a highly artificial measure, contradicting the very nature of mobile media, as the empowering part of mobile media is precisely its independence from locality.[6] The trajectory is a correlation between registered citizens, invested with rights, and mobile media use. As such it is an re-instantiation of the sovereign-governed relation in a field that in its historic roll out was a society-homo economicus relation. That the state and bureaucratic institutions now inscribes themselves into this field shifts, the mobile media regime from solely economical factors to that of citizenry, or at least residency. By differentiating and discriminating the population by criteria such as proper residental documents, it targets at social segregation and exclusion.

> Mobile phones is a most fundamental thing, because when you do an occupation, there are no landlines. But you need support on different levels: political, every day, food. A land occupation is very precarious. When you do an occupation you need to build a lot of networks to sustain the occupation on all different levels. The only way to do that is via mobile phones.
>
> Fernão / São Paulo

Hardlinking proof of residency and mobile media inclusion re-enacts not only a segregation along social factors, but can as well be understood as a translation of borders on a transversal level into the interior of a country: a correlation of spacial borders to communication borders.[7]

The second illiberal prospect of mandatory registration shows some parallels to data retention, as in addition to social sorting it produces *the possibility* of illiberalization by setting up an agglomerated database which connects SIM card data and personal data. The linking of these two distinct data sets provides firstly a factual identity between the device and its user. Practices such as lending it to friends for making a call or sharing one device within a community are being delegitimized, and at the same time they provoke irritation to the government scheme.

> Txtpower has been opposing the registration of SIM cards, here in the Philippines you can buy them still anonymously. Our members fear the possibilty having our data compiled by the government. That could lead to surveillance. There were campaigns, as well as against the national ID system
>
> Eder / Manila

6 In contrast to a registration for landlines.

7 Some countries, e.g. Turkey, let foreign SIM cards function only for a limited time.

Prepaid SIM cards outnumber postpaid cards internationally by far. Mandatory registration is not addressing a small business, but the by far biggest part of mobile phone use.

> The prepaid is dominant, 95 to 99%. It gives a certain level of anonymity, especially in the current situation where human rights are deterriorating.
>
> Alegre / Manila

Due to the unavailability of banking accounts for the majority of the world's population, cash paid telephony was a necessary condition for the huge success of mobile phone use in poorer regions. In fact, without unregistered prepaid cards, mobile phones would not play the communication role it is actually playing today on a global scale. It is a seldom observed historical process to diminish the user base of a technology after its massive roll out by inventing measures that exclude many from its use. The prospects are high that this anti-liberal rollback produces frictions with governmentality programs, and it remains to be seen what solution emerges out of these frictions.

Additionally, it criss-crosses mobile payment strategies. To be able to use the mobile phone for the transfer of small amounts of money was an answer to the lack of access to banking accounts for large parts of the population. One of the reason banking accounts were inaccessible was lack of proper documents, most importantly of residence. When the mobile phone could bridge this lack, as it would be used as a mobile residency, a mobile society gets rebound to a territory of a sovereignty via registration of SIM cards. In this sense, this measure is pure sovereignty: reterritorialzing a liberalized scheme, invoking a territory.

Blocking of Communication

> There were many efforts to smuggle cell phones into the house of the Chief Justice during his house arrest. And then they just set up phone jammers around his house. In that sense, there was active jamming.
>
> Theriot-Orr / Seattle

Mobile media invokes governmentalization of the population. The blocking of communications then interrupts governmental rule by policing. As Opitz notes, "the police is a limit-figure actualizing a sovereign relation: it is a spectral force that contaminates the law in mundane governmental practice" (2010, 107), a practice, which only post-factum is subject of juridical inquiry, if at all.

To block contents on the internet has become a usual practice for regimes. It might be drastic, as in China, or as in Bangladesh, as Swapan reports:

> The government announced last week that 84 porn websites will be blocked, means that it will be investigated. They are surveilling and finally notice this. There is a blog in bangla that was blocked, because it published a picture of a war criminal and the blogger was blocked. It could be accessed from outside the country. YouTube was blocked last month. A talk between politicians was put there. People got to know because of the blocking. I talked to the president of the ISP associations and he said they are blocking, filtering, because the government tells us. There are also sort of house to house raids. One blogger was arrested, beaten severly. He left the countries and now lives in Sweden. Because of the expression of his political views. So, there is much more than we know happening
>
> Swapan / Dhaka

The limitation of access to website is a common topic in the discourse of freedom of speech. Less known is the political technology of blocking mobile cells, single phones, or segments of a network.

It is hard to analyze the reasons for suspicious symptoms of mobile phones: there might be network problems in general, or other unintended ruptures. Francisco from Mexico City tells about different effects that he identifies with blocking communications.

> When we ring a phone, it sounds busy, and the next time it goes through. They also block certain numbers, from my phone I can not contact specific numbers, another thing, the phone does not ring for hours and says that it is out of the area of coverage when it is inside. In regard to messages, one thing is that when I send messages, it says it is not sent, but the other person received it and the other way round. Sometimes it says it was sent, but it never reached the recipient, also message sometimes take two or three hours to reach.
>
> Francisco / Mexico City

Whether this is intended or not remains opaque. But as many activists reported about similar problems, one might accept the fact that targeted blocking of communication is a reality for some activists.

In this section I want to focus on communication blockage that does not target a specific phone, but a specific location during protests and demonstrations, such as what Hamid Zaman refers to.[8]

> At other times what happened was that the entire phone system would be jammed for a few hours, we couldn't send messages or even call people. When something big would happen they didn't want the information to get out. I think several times for a few hours the networks were blocked
>
> Zaman / Lahore

The BART Case

To study mobile network shutdowns one does not need to look at Pakistan or Egypt. San Francisco had a similar case. Here, a police-sovereign intervention of this kind does not necessarily need to be invoked by police as such, as the case of the metropolitan transit system in the San Francisco Bay area, known as BART, shows.

On July 3, 2011 a transit police officer shot a man on the premises of BART[9], which spurred protests. Some of which where planned online, so that

> BART officials decided to disable communications by powering down the mobile service repeaters/microcells and the Wi-Fi routers in order to inhibit this planned on-line coordination. The protest never happened, but all transit users suffered from the loss of connectivity.[10]

This shutdown of communications was criticized widely, amongst others by the American Civil Liberties Union (ACLU) and the Electronic Frontier Foundation (EFF). BART claimed that there had been a threat to public safety, thus allowing them to shutdown connections. At that time, BART did not have a policy for such an operation. A policy was only introduced after the fact.

---

8 Here, he reports about one incident during a manifestation in Lahore.

9 "Man shot to death by BART officer identified", Kevin Fagan, San Francisco Chronicle, 8 July 2011, http://www.sfgate.com/cgi-bin/article.cgi?f=/c/a/2011/07/07/BA9U1K7O2C.DTL (accessed 4 March 2012).

10 "Statement on San Francisco BART cellphone service shutdown", EFF Austin, http://effaustin.org/2011/08/statement-on-san-francisco-bart-cellphone-service-shutdown/ (accessed 4 March 2012).

The availability of mobile communication in the metropolitan transport system was introduced after 9/11 "in response to popular demand from BART riders who saw that New Yorkers had found cell phones to be invaluable to communicating with authorities and loved ones in the midst of city-wide confusion."[11]

In the BART case clearly "a form of government follows the logic of police if it transcends the law by applying it. In the name of public order and security the law is violated and, through this act, reestablished" (Opitz 2010, 107). The reestablishment in the BART case was disputed by Civil Liberties Groups, which led to a policy that made the suspension of communication legal for severe cases of threats to public order. When such a situation appears, remains opaque. Protests by their very nature try to disturb public order as a means to gain momentum and public attention.

Police Interventions and Mobile Phone Blocking

Activists in general, during street struggles for social justice, or even in the preparation and organization of those, regularly face a situation, that is generically referred to as a "police state". Whereas "police" refers to very different entities and different kinds, such as undercover, anti-terror units or intelligence units, depending on the 'tradition' of sovereign rule in each specific area and the situation.

The logic itself though remains the same: a limit-figure enters the arena in a different mode of relation towards the population, a mode of temporal, situation-bound sovereignty, which utilizes force and counter-communications means. These operations are often not performed on the grounds of lawful activity. Activists in such situations have no other option but to run a juridical case against the police activities later on (and often win). The situation itself, though, is *contaminated via sovereign rule* and the right of public gatherings and free speech repressed.

Blocking of mobile communications is a tactical measure to weaken the activist's potential of orderly collective actions. Mobile media seen from this angle highlights the costs of its use for activists: once integrated into the activist's agency, its unusability produces a void that cannot be refilled ad-hoc.

> Your are sitting down and police surrounds you, they are not seeing you are texting. Before they take you with them you at least send

[11] "Want Public Safety? Don't Disable Cell Phones", Eva Galperin (EFF), https://www.eff.org/deeplinks/2011/08/want-public-safety-dont-disable-cell-phones (accessed 4 March 2012).

> a message out. And that has helped us. It happened just one week ago. We were at the landfill and a gang of Ghundars, Hooligans, they are bought by the city authorities to basically surround us and beat us up, were brought in to attack us. And we were working there with the communities who are affected by the pollution. So the only thing I could do was text and the text helped to get a lot of the press involved so fast, that the press started to call senior authorities and the senior authorities rushed in. So, it protected us to text.
>
> Saldanha / Bangalore

In a recent comparative case analysis of the occasions in which regimes disconnected infrastructures at large, the authors "demonstrate the importance of understanding how information technologies have a role in political responses and counter-insurgency tactics of many kinds of regimes" (Howard, Agarwal, and Hussain 2011, 221), as it shows that "that there are a range of occasions in which state power over digital networks is noticeably strong" although turning off the networks in Egypt drew out the protesters to the streets (Howard et al. 2011, 220), which demonstrates very drastically the failing of this governing measure. While their study remains a bit unclear[12], it generally shows that almost all kinds of state interference with digital networks happen in formal democracies as in so called authoritarian or repressive regimes alike. This underlines that, for a study of governmentality at least, the perspective on formal designs of government is neglectable.

M¡ point here is: a political technology is temporarily annihilated in the stage of neo-sovereign rule. But this does not mean that it stops performing. It just does so in different ways. The sudden breakdown of a resource which is deeply integrated into the activists' practice targets the very agency of protest partakers. Additionally, it isolates them, as it becomes hard to contact lawyers or medical aid.

> Since WTO 2005 in Hong Kong, lots of my friends always suspect that our mobile phones have been surveilled for many years already. Because if something happened, people gathered to protest or assembly on the street, our phones were cut off suddenly. Of course we hardly have evidence that our phones have been surveilled.
>
> Freddie / Hong Kong

---

12 To count the 1995 blocking of ISPs in Germany amongst state interference in public access to *social* media calls for a more precise distinction of digital media types.

Data Retention as Neo-Sovereign Data Production

"The biggest use of wiretapping tools is not actually the capture of conversation but something that is not really wiretapping at all: the capture of transactional information." (Landau 2010, 99) Susan Landau opens the theme by pointing to the ambivalent status of data retention: it is not wiretapping proper, but still using the same tools.

The European Union Directive on Data Retention,[13] though less than 10 pages long, is invested with considerable authority. It directs the member states to pass a law compelling each provider of telecommunications services to retain traffic/location data for at least the past six, and at most, the last 24 months. As stated in the first sentence of Article 1:

> This Directive aims to harmonise Member States' provisions concerning the obligations of the providers of publicly available electronic communications services or of public communications networks with respect to the retention of certain data which are generated or processed by them, in order to ensure that the data are available for the purpose of the investigation, detection, and prosecution of serious crime, as defined by each Member State in its national law. (European Commission 2006, 56)

"Certain" data here means traffic/location data, and as defined here, is data generated by or during an act of telecommunications with a mobile phone, a landline, or via the Internet, minus the "content." These inquiries around data ask who, when, where, with whom, how long, and so forth — but do not ask about the nature of the communication. The data generated during unsuccessful acts of telecommunication is also similarly analyzed.

> To protect the secrecy, people are encouraged to use the telephone. The email is written down, so it is very easy to be surveilled, but voice phone is only caught by wiretapping, which is rare.
>
> Yasuda / Tokyo

13 European Union Directive 2006/24/EC of March 2006 on the retention of data generated or processed in connection with the provision of publicly available electronic communications services or of public communication networks and amending directive 2002/58/EC, hereafter: 2006/24/EC.

To "harmonize" means to implement technical standards of retention, and to do so for data access from anywhere in the EU.[14] The data one profiler gets from a member country shall technically be compatible with the data she obtains from any other member state. Thus, corporations that provide such technologies are forced to engage in standardization and interoperability.

Unquestionably, the directive rests on a differentiation between traffic/location data and content data. The retention of all data generated during an act of telecommunication assumedly does not fit with directive 2002/58/EC on data processing and privacy of 12 July 2002, and other fundamental human rights.[15]

This differentiation between data that contains the structural components of communication and data that relates to content is, first of all, technically inspired, and thus indicates the possibility that the directive is produced with this technical differentiation in mind. Traffic data consists only of the information needed to technically initiate, sustain, and terminate an act of communication.

The relevant data is divided into the following subcategories:

- data necessary to trace and identify the source of a communication;
- data necessary to identify the destination of a communication;
- data necessary to identify the date, time, and duration of a communication;
- data necessary to identify the type of communication;
- data necessary to identify the users' communication equipment or what purports to be their equipment; and
- data necessary to identify the location of mobile communication equipment.

---

14 It is worth mentioning that this directive has passed as a so-called First Pillar directive using the single market power of the European Union, and not as a Third Pillar directive pursuant to the Union's power fighting crimes. "Once the choice was made to go ahead with the Directive as a First Pillar initiative, the Commission and the Council took the position that, legally speaking, the Directive could not regulate police access to communications data. Anything having to do with the police was strictly Third Pillar." (Bignami 2007, 12)

15 If the actual implementation does so, and if it complies with national laws on privacy, is not of concern here, and neither is the discourse on data retention and citizen rights. For a historical discussion of data retention in Europe and Canada, see Warner 2005.

The term traffic data itself is problematic here. Article 6, sentence 2, of the directive states: "No data revealing the content of the communication may be retained pursuant to this directive." But the person calling the crisis line is not ordering pizza. There are undeniably semantic elements in the mass of so-called traffic data.

> For the shelter we don't use mobile phones or landline. The shelter is a secret place, so the meta data itself is a problem.
>
> Raj and Kannan / Bangalore

Data identifying the person calling the crisis line unavoidably reveals content by virtue of the transaction data itself.

Data retention, as laid out by the European Union in its directive 2006/24/EC, seen from its angle of *how* it works, can be divided into two parts: first, the collection of transactional communication data of the whole population. Second, when such data becomes operational within police or secret service investigations.

Data retention is a measure that is invoked without targeting concrete persons. Thus it is a strategic surveillance scheme, not an ad-hoc, tactical one, like the deployment of IMSI catchers.[16]

I have argued elsewhere (Leistert 2008), that data retention of meta data changes the 'ontological' state of such data: in the first place this data is ephemeral and a necessity for telecommunications: it is needed e.g. to connect person A with person B. By storing it, this operational, ephemeral data ontologically changes into second order information as it now provides access to the history of telecommunication of individuals and networks of people.

Retaining traffic data shifts the data regime from pure machinic to human, from operational addressing to representational referencing.

While transaction generated data is generated during a telecommunication act, its *signification* outside the processing computers that belong to the telecommunications infrastructure itself is produced *by its retention*, via standardized access and, later, its reference to people's names. Retention is a modal operator as it shifts

---

16 IMSI catchers are tools that simulate a cell. The mobile logs into this trap, and connects through the catcher to the real network. All communication to and from the phone then can be tapped with the catcher. Why this is possible, see section 9.

registers of data. By retaining the data for the purpose of crime investigation, a significant shift transforms the prior technically necessary data into data that now has a meaning for humans, and so is strictly speaking *content data*. The process of retention itself inherently supports the transformation of data. Now, it is archived on some dedicated storage media in a dedicated storage form, easily accessible and search-able.

Taking this shift seriously, one may speak here of *new* data and reject the common notion of an identity of traffic data and retained traffic data. The retained data has no computational function anymore, but it is transformed into the realm of the symbolic: It now *represents* the movements and telecommunication acts of people whereas previously it was not placed in the register of representation at all.

As traffic data is produced automatically by communication technology itself as a working necessity, the production of the profiler's representational data is done via copying transaction generated data into a database.

> The police is trying to understand the organisations, e.g. which group is led by whom. And that this group is close to this group. So they are mapping the whole thing. The contents of email is not so important, but it is much more interesting for them who is communicating with whom.
>
> Yasuda / Tokyo

This transformation of traffic and location data into content or representational data cannot be described within the logic of the act of communication itself, as this act is outperformed by machinic operations. The database containing former meta data is a *newly generated object*, produced by specific algorithmic operations and strategic settings.

At first sight this is in accordance with political technologies of governmental rule, as "knowing an object in such a way that it can be governed [...] requires the invention of procedures of notation, ways of collection and presenting statistics, the transportation of these to centers where calculations and judgments can be made" (Miller and Rose 2008, 30). By way of collecting the meta-data of the population, an object is, produced and made intelligible. Data retention can be seen as a productive endeavor, it not only allows access to transactional communication data, but first of all, by intervening in the temporality of meta-data, it makes it persist in time. Data retention thus is data production. It harvests this data. Such cases, where data that is used for other purposes than its initial use is called *function creep*. "[I]nformation gathered for one purpose can easily be reconfigured for other uses, thanks to the ease of copying, storing, searching, and transporting digital data" (Andrejevic 2007, 176; see also Winner 1977). In the case of data retention, the specificity of this function creep is that the data initially was not even stored. In that sense, the process maybe better termed "production creep".

> You see the impossibility that this data is used for anything reasonable. London is one of the cities with the highest crime rate in Europe. The means of the surveillance are mainly political, not practical, there were lots of weird examples of surveillance when lots of the councils used the terrorism act to use cameras to track pupils who didn't go to school.
>
> Startx / London

One as well should remember a general aspect of large technological materializations, as they

> produce unexpected problems, are utilized for their own ends by those who are supposed merely to operate them, are hampered by underfunding, professional rivalries and the impossibility of producing the technical conditions that would make them work – reliable statistics, efficient communication systems, clear lines of command. (Miller and Rose 2008, 71)

In a broader context, Lyon refers to deregulation as it

> is contributing to the blurring of boundaries between sectors that handle personal data. The technological availability factor works with not only the desire of government and commercial corporations to increase their data handling scope and efficiency, but with the devolution of government data processing to private organizations. (Lyon 2001, 51)

While this might mean that the systems do not work with the effectiveness the policy makers have been looking for (in regard to the harmonization pillar), a function creep might occur at many more places before this data is transferred to possible second use databases: 27 member states offer their police forces, secret services and other agencies access to a distributed data base ecology, which is administered by the telecommunication providers themselves – from Spain to Romania and from Finland to Cyprus.

For Mexico, Blax reports about the common problem of a function creep of data intercepted by surveillance:

> Fraud is the most common use of surveillance in Mexico. There is a whole industry of corrupt buerocrats, banks, and others that collect your data and try to get money by doing fraud. Up to simulating the voice of some of your relatives that then calls you and ask for money
>
> Blax / Oaxaca (Mx)

I was told similar stories in many other countries. The opaque collection of personal data and data linking to persons empowers a diversity of legal and illegal activities. Access to the data collections will never be controlled in such a way that function creep becomes unlikely. Function creep is not only a problem of bad data security or others deficits, but has its systemic place in a networked and computerized society.

First Effect: Disciplining Users via Fear

2006/24/EC already has direct effects on the level of retention/production, without that the data is used for anything: specific groups of professionals, like journalists, priests, and lawyers, whose work depends on confidentiality to their clients, have expressed their concerns as they fear their work will be influenced, or even made impossible, if they can no longer provide confidentiality.[17] For a technology that operates within a governmental society, this is a problem, as it must not intervene into the economic activities of the subjects (for example the business of being a lawyer).

In general, cautious individuals may change their communication behaviors, may stop communicating with real or imagined, current or future targets of surveillance. This is another level of direct intervention where a liberal apparatus of security is rendered into illiberality: the production of fear, up to the point where communication stops. In this field of operation, data retention is part of disciplinary rule. It prescribes the individual to disintegrate from 'dangerous' or 'suspicious' subjects. The level of threshold is highly speculative, as the decision to

17 On the controversy about data retention in Germany, see Meister (2008).

consider someone 'suspicious' cannot be made with direct reference to that person. The virtual possibility that this person, or better this person's digital double, might encounter in the modalities of securitization is discipline by imagination of an opaque assessment. The individual needs to make a risk assessment towards her social network, discriminating amongst her friends and contacts who might be a risk. The temporality plays a pivotal role: it is simply pure speculation to anticipate a future investigation into an acquaintance. "Expertise in all spheres of social practice is redoubled against the background of a perpetual threat", Mark Andrejevic writes in his analysis of post-9/11 policy in the US (2007, 165). This effect of "generalized chilling" posits "a maybe state. Here, a homogeneous social body is produced by virtue of what lies outside of it" (Fuller 2007, 146). Data retention on the level of its basic operation (collecting the data) already is an invocation of an algorithmic neo-sovereign rule, although it does not limit communication as such. It shares a common aspect of surveillance technologies, the more if they are unsharp, broad, and rather unspecified: a normalization effect.

Hannelore Bublitz refers to the mechanisms of control as a quantification of control mechanisms that alter into an automated structure, which leads to a coercion to adjust to this conformity, that on the level of population "via the homogenization, adjustment, and normalization of the plurality of masses provides normality and security" (2010b, 162; translation O.L.).

In the context of CCTV this has been largely documented and debated. Surveillance and dataveillance have direct effects on behavior of unspecified subjects. The discipline modality urges the subjectivity to adjust in a self-regulating manner. Normalization here is invoked via the instrument of fear.[18] Neo-sovereign fear is not caused by contents or behavior, but because it oversees the contacts, the social network. Not actively, but in the sense of the panspectron.

18 Brian Massumi discusses fear as an instrument of rule in the post 9/11 events explicitly by referring to the code warning system that is ubiquitously present throughout important public places in the US (2005; see also 2009).

Second Effect: the Materialized Unspecific Doubt

We are living in a democracy. But who knows about tomorrow?

Alface / São Paulo

Understanding data retention as a technology of illiberality within the general liberal scheme of mobile media or internet communication marks its specificity: data retention does not block communication flows of the population. It does not disturb productivity and circulation in the course of its operation. Much more, it *takes advantage of the flow of signs* initiated by liberal rule. But while data retention works on the whole population by default, it invokes a different relation towards the population than liberal rule: the population now returns as a risk. In a sense, data retention is the materialized doubt about civil society as self-regulating. Illiberal governmentality invokes securitization, to secure its apparatuses of security. As Andrejevic in the context of the "Total Information Awareness Office" explains, and I argue the same can be said in terms of EU data retention,

> the data gathering process is a fishing expedition, designed to generate suspects by sifting through the data and identifying potentially high-risk individuals. Data gathered about the rest of the populace is essential to the process: it helps set a baseline pattern of behavior for investigators, one from which any evidence of deviance triggers suspicion and further investigation. (Andrejevic 2007, 177)

It opens the possibility of statistical analysis of communication in regard to differentiated normalities, as it offers the proliferation of social groups. To be able to differentiate on a statistical level, i.e. to discriminate subsets, data retention provides necessary elements for neo-sovereign rule: based on patterns that are algorithmically produced, the ground is laid for the machinic production of suspects. "As any intelligence operative knows, communication patterns are full of information. From them one can discern organizational structure as well as behavioral patterns." (Landau 2010, 89) The axis this algorithmicity works on is temporal, as traffic data is collected over time. Of course, locational data is collected as well. Here the designated surveillance space is the European Union. This is rather natural, as it is an EU directive. But the specification of the temporality needs to be addressed by the directive specifically, 6 to 24 months.

As an example of the current realities of this directive, I refer to Poland: the directive had passed parliament without much public discussion; now recent numbers show extreme use of this scheme as "in 2011, Polish authorities requested users' traffic data retained by telcos and ISPs over 1.85 million times—half a million times more than in 2010." This figure can be seen as an effect of almost unlimited use cases as the

> law allows authorities to use the retained data in an almost limitless range of scenarios, including petty civil offenses and minor criminal investigations. Moreover, Polish authorities—ranging from law enforcement to intelligence agencies—can access the retained data without independent oversight and at no cost. Rather than require authorization from an independent judge, the law permits access to the data through a simple written or oral request authorized by the head of the Central Anti-Corruption Bureau, the Polish intelligence agency on anticorruption.[19]

It is this number, 1.85 million requests, which signifies the qualitative leap of this technology. Although one instance remains the single user or surveilled citizen, that, for whatever reasons, triggered a request. The other instance, and this is both connected to the first and points way beyond, realizes itself within 1.85 million requests. This sea of relational data about who communicates with whom echoes other political algorithmic technologies, such as the troublemaker database, in one trajectory: it provides the realities of algorithmic performative neo-sovereign rule. It is this data that *produces reality* from which individuals are deduced in the first place. Not a dragnet investigation in the strict sense, the operations that are rationalizable by data retention still provide a new episteme of rule, which is highly performative. Such "an informational space is inherently immersive, excessive, and dynamic: one cannot simply observe it, but becomes almost unwittingly overpowered by it." Terranova writes about the informational milieu which she sees as the base of politics in the information age. And she continues: "it is not so much a three-dimensional, perspectival space where subjects carry out

19 "European Data Retention Directive At Work: Polish Authorities Abuse Access to Users' Data" by Katitza Rodriguez, EFF. See https://www.eff.org/deeplinks/2012/04/european-data-retention-directive-work-polish-authorities-abuse-access-data (accessed 12 April 2012).

actions and relate to each other, but a field of displacements, mutations and movements that do not support the actions of a subject, but decompose it, recompose it and carry it along" (Terranova 2004, 37).

Data retention thus is a materialization of a liberal excess that is nurtured by (amongst other technologies) mobile media as it invokes a general mass surveillance media.

Third Effect: A New Knowledge - Power Axis

> IT is the solution. Because we have millions of millions of people. They have central database and can connect everything. And later do data mining. It is a solution for the fear they have. If there is a problem, they can select on who is good and bad by only picking this abstract data, without reflecting the feeling that you have. You can be happy or angry when you do something. It is like when you take some information from people with torture, you can not be sure if that is true, because people under torture say anything.
>
> Iokese / Madrid

A data retention scheme on communication datahas the effect that its outcome is indisputable for the objects/subjects: the system produces a reality. The process of how this match between a record in the scheme becomes a reality for the object is blurred: there is no informed consent from users of mobile phones to collect the data. Just like the diagnosis of a disease, whose existence had to be standardized in a scheme to become true, the appearance of a user's location data from 6 months ago provides an incontestable truth, while its production is abstract and out of the hands: like an all overseeing sovereign, data retention tells the truth over a territory and its subjects. Only that these subjects are made up from flecks of identity which are processed over and over again.

Data retention thus addresses and reshapes the present of subjects by confronting it with the subject's past represented in the subjects' *doppelgänger*. As such, it algorithmically reverses time by disqualifying the present in favor of the past. Communication done a year ago can now produce a deviant digital double of the present. And further, in pre-emptive modulations of police work, predictions about the future are easily computed.

Once a suspect is identified in the pool of data, all contacts of that suspect are reconstructed over time; patterns, recurrences, locations, behaviors, modes of communications. If this pool of data is analyzed within an §129a investigation (see chapter 8), investigators have the right to claim all contacts as suspects as well. This is where data retention suddenly turns productive in a different sense as it *generates suspects*.

Production of statistical anomalies as a foundation of discrimination is the liberal process reversed, as the aim is then not to work on these isolated subgroups via governmental programs that aim at conducting to self-conduct. The isolated subgroups are discriminated only to be subjected to neo-sovereign measures. Whereas liberal statistical knowledge genuinely starts from the assumption of a broad field of tolerable deviances as statistically normal, and only later defines a threshold beyond which it assumes problematic zones, data retention as a program turns this logic upside down. The data is only and solely used by police and secret services to enhance means of investigations. The effect is that only a very small proportion of retained data directly relates to what the directive was introduced for: the fight against the serious criminal (or better: communication acts that are triggered as relating to serious crime), while billions of other data sets relate to untargeted communication acts.[20] Data retention supplants the history of the individual "as a socio algorithmic process. Flecks of identity are variables and events. From the perspectival scale of such systems, life is a trail of triggers and tokens: date, time, location, status, speed, choice, amount, accomplices." (Fuller 2007, 149).

---

20 This huge disproportion has severely alarmed privacy commissioners and civil rights organizations alike. The former mostly argue that the idea of proportionality is abandoned with data retention. The latter fear that once the data is there, it will be used for other projects as well, most prominently to chase users that are suspected by the cultural industry to infringe copyrights. The value of this data collection, of course, is the next issue. Data mining companies, that already hoard many datasets on large parts of the population from commercial sources, like *ChoicePoint* or *LexisNexis*, would get access to a whole new knowledge about behaviors and social relation. Just as an example: Blax reports that the US company ChoicePoint was active in Mexico:

> Choicepoint purchased the data of the main photography identifications, it is a electoral card. Choicepoint now has the adresses, pictures, fingerprint, your signature... That was in 2003
>
> Blax / Oaxaca (Mx)

The *Washington Post* reporter Robert O'Harrow documented in his book "No Place to Hide" the amalgamation of government and commercial databases in the post-9/11 events in the USA with a detailed description of the leading companies involved (2006).

The rationality of the database provides the defining sources to enforce the liberal paradox of illiberality. It appears independent from personal or subjective informed knowledge. This data is enacted as true, as a truth, because it provokes practice and activity which grounds itself within the horizon established by these political technologies of securitization. One might call this technique *algorithmicly produced sovereign performativity*.

A Panspectron Supplements Liberal Technologies

The rahter conservative Royal Academy of Engineering (UK), in respect to huge increases of surveillance practices by the state, addresses necessary conditions for the application of surveillance technologies:

> Reciprocity between subject and controller is essential to ensure that data collection and surveillance technologies are used in a fair way. Reciprocity is the establishment of two-way communication and genuine dialogue, and is key to making surveillance acceptable to citizens. (Royal Academy of Engineering 2007, 8)

In light of the operations undergoing permanently by a panspectron, such claims seem anachronistic. Technology has widely left behind panoptical settings, which on the phenomenal side are often visible (such as CCTV). A parallel in time remains, though. The panopticon and the panspectron are historically at least coinciding. Data retention of telecommunications data as well as 'troublemaker' databases, or the uncountable commercial data collecting entities, that track which websites one views and what book one buys, belong to an assemblage of new emerging forms of control endemic to a society that relies on networking infrastructes. Its power results from the wide acceptance and usage that electronic communications media have gained in the last years.

Surveillance of mobile communication is nothing exceptional: data retention served as an example for mass surveillance in this chapter. To make visible social networks is one of its powers. Nathan Freitas tells this episode of a more sophisticated communication circuit, which counters the surveillance on social networks by constructing a different way to route the connections:

> Finally I have done some work with Voip, Asterisk, Sip. Redirects and call-ins that bounce you to another call, hide your phone num-

> ber. It is still emerging. We use that in Beijing. We had seventy people go there, each person had a different phone number, virtual number that they would call into. To check in. So you could not say that this person and this person are calling the same people. I was able to provision 70 different phone numbers in different states and countries. So that no one can be linked together.
>
> Freitas / NYC

But this is a temporary set up. What remains is that calling someone produces data far beyond the call. And while the call might be forgotten to have taken place within a couple of days by the people who have spoken to one another, the technical infrastructure implemented for data retention ensures that, for up to 24 months, it happened, regardless of whether or not any human being remembers the cal, because the database does so upon query.

8 Mobile Surveillance Media Case Study

> We try to make sure the communication between our users is safe and our server is safe. We try every day to make the system safe. We are thinking about the importance to be protected against these measures. The social movements in Mexico are very unprotected. They need to learn from each other to gain more safety. Because as we can see, the situation might become worse. In a technical perspective a lot of servers are unprotected, a lot of people use their mobiles without securing it. People are sending very important information through mobile phones. Names and places. They think that mobile phones are a very special media, that no one can see the messages. It is very important I think to make them conscious about it, a lot of very active people don't know anything about it
>
> Olinca / Mexico City

To provide an account on the contemporary possibilities and practices of authorities to investigate and do surveillance tasks with the help of telecommunication providers, the case of the German "Militante Gruppe" (MG) is insightful. It offers a perspective on what law enforcement agencies (LEA) in practice do and how their inner logic perpetuates these operations.

The material shown here originates from preliminary proceedings and had been delivered to the courts by the investigating authorities themselves. This makes them highly valuable in a specific sense: usually it is only possible to interpret symptoms of surveillance or be faced with the consequences, whereas here a documented account of such measures is available.[1] What conclusions the authorities drew from this material is not of first interest. Much more central here is what kind of surveillance had been used, how data had been received, what data was

---

1 The material documented here is provided by sources that prefer to remain anonymous. These are extracts of scanned paper pages produced by LEAs. The court reference numbers are GBA 2 BJs 58/06-2 and ST 45 / ST 14 – 140011/06.

specifically produced to surveil and so forth. For the subject of cybersurveillance only technical surveillance aimed at the telecommunication of the accused is represented here. Left out are all other kinds of surveillance, such as persona or video camera surveillance, which was also conducted during the period. What is more, details of eavesdropping operations of phone calls are ommitted. Suffice to say that all calls to and from the accused's mobile and fixed-line phones had been intercepted, as well as those of relatives and friends. The number of those affected by surveillance operations adds up to more than 200, even though there were only three suspects.

The primary purpose of presenting this material is to demonstrate the weakness of telecommunication infrastructure with regards to privacy. This example is illustrative as no legal obstacles prevented the surveilling parties from their work in any way due to the application of the anti-terror law §129a. These surveillance operations are presented here on a level of documented evidence and demonstrate, in a nutshell, the technical possibilities of telecommunication surveillance unleashed.

In the context of an analytics of government, these practices relate to a neo-sovereignty that rules from a distance, over infrastructures that governmentality had laid out before.

Introduction: Section 129 of the German Criminal Code

Some context is needed beforehand: in Germany, law enforcements' legal means for infiltrating, surveying and detaining political opponents after World War II have been steadily extended since the 1970's. The most prominent case is §129a of the criminal code, dealing with terrorist organizations. Its older variant §129 deals with criminal organizations and actually predates the Federal Republic of Germany. It was extensively used during the Nazi era for the persecution of imagined or real opponents. Only a few years later, in the 1950s, hundreds of investigations against alleged communists and activists opposing the rearmament of West Germany were performed applying this controversial paragraph. Subsequently §129a was introduced, covering terrorist organizations in the 1970s, and after 9/11 §129b was introduced, dealing with foreign terrorist organizations.[2] These laws basically

2 An official English translation of the German criminal code incl. §129 can be found at http://www.gesetze-im-internet.de/englisch_stgb/englisch_stgb.html (accessed 4 March 2012). But the means provided are defined in the code of criminal procedure (Strafprozessordnung (StPO)).

strip suspects from every last bit of their (privacy) rights and have been used effectively by LEAs most frequently to update their knowledge on leftist activists. Hardly any of the numerous §129a investigations made it to court. More than 95% are silently shut down, often after years of very intense surveillance.[3]

The introduction of this paragraph in the 1970s was, amongst other means, meant to deal with the Red Army Fraction (RAF) and the Revolutionary Cells (RZ).[4] Although these actors now belong to the past, the paragraph has continued to see numerous applications to investigate leftist or social justice groups.[5] The application of §129(a,b) allows far-reaching surveillance not only of those under suspicion, but also of those who have been in contact with those under suspicion, even if only once. As the suspects are commonly engaged in a diversity of political fields, the application of such an investigation produces a complex reproduction of the social net of many politically active people, regardless if they themselves are suspects defined under the investigation or not. Sharing a flat, belonging to the family or working at the same company is enough to become a target of extensive surveillance once §129(a,b) is at work.

Especially StPO §110a (surveillance of telephone and post), §100c and §163 (long-ongoing observations), §110a, §110c (systematic deploying of undercover agents and spies), and again §100c (surveillance of acoustics and images inside private homes) give almost unrestricted powers to LEAs. An English translation of the StPO can be found at http://www.gesetze-im-internet.de/englisch_stpo/index.html (accessed 4 March 2012).

3 The surveillance conducted under these paragraphs does not always seek to remain unnoticed. The suspects thus react to the investigations and provide the LEAs further insights into their social net. Additionally, suspects reportedly have lost jobs and suffered as well psychologically from investigations. As such this is highly problematic as this all happens even before pre-trial confinement.

4 Both groups were using force to achieve political goals, the big difference is that the RAF, whose founding generation is known as Baader-Meinhof-Group, went underground whereas the RZ personnel had their normal day job while pursuing RZ activities 'at night'. Their genealogy can be traced back to the events of 1968, in part even until the protest against the rearmament of West-Germany in 1955, which was a big debate at the time.

5 Right-wing groups are very rarely targets of §129(a,b) although violence originating from right wing groups in Germany, especially against humans, including murder, continues since Germany's reunification.

The MG Investigations

The MG investigations, into the alleged terrorist group known by these initials, started in 2001 (with some pre-proceedings by the German secret service since 1998). Amongst the many different allegations were attacks on German military equipment. The material shown here is from the so called "MG 1" investigation, which started in 2001 and ended in 2008. Others are still pending, as there have been numerous different investigations.[6] The surveilled had been accused of forming a terrorist group. None of the accused in this case have been sentenced and the case has never become a regular court case.[7] Additionally, on 11 March 2010, the Federal Supreme Court (BGH) ruled that the entire set of procedures used by the LEAs in this case (MG1), had been unlawful and that the surveillance conducted was not appropriate, as there was never enough reasonable suspicion.[8] The data collected nonetheless remains in the police archives and has recently been transferred to the Berlin criminal state police.

In the following passages, some details about the surveillance operations are explained. They should be seen as examples from a myriad of others. The documents covering this "MG1" investigation come to thousands.

Retained Meta Data of (Mobile)-Telephony

Figure 8.1 shows a typical request for information [*Auskunftsersuchen*], which reports call data from 1.10.2006 to 31.3.2007 (only 3.10. to 19.10.2010 is shown here, but the duration of 5 months is mentioned in the upper left area) from one MSISDN (phone number). Besides the number dialed, other information is printed: Mobile Country Code (MCC)[9] 266 refers to Germany, Mobile Network Code

6 Details about the cases, their time lines and each investigation are published at http://einstellung.so36.net (accessed 4 March 2012).

7 Anne Roth, partner of one suspect in these investigations, Andrej Holm, has been blogging extensively about her life under total surveillance. Here, details are published, partly in English: http://annalist.noblogs.org (accessed 4 March 2012).

8 The rule is online (German only): http://juris.bundesgerichtshof.de/cgi-bin/rechtsprechung/document.py?Gericht=bgh&Art=en&sid=52c1c2b856536c08dab95908724bccfd&nr=52160&pos=0&anz=1 (accessed 4 March 2012).

9 A code defined by the ITU to identify countries of mobile phones. It is a part of the IMSI.

(MNC)[10] 01 refers to T-Mobile, the Mobile Switching Center (MSC)-ID[11] is responsible for the end-to-end connection, e.g. allowing hand-over requirements during the call. The cell-ID references the actual cell where the phone was logged in, and on the far right, geographical positions of this cell-ID are printed. Generally, all telephone communication meta data to and from the mobile phones of the suspects had been retained and provided to the authorities. These meta data include, amongst other rather pure technical parameters, the phone numbers, the IMSI[12] (if applicable), duration of call, type of call (it differentiates between service call types), and the geo-coordinates of the cell where the client was connected. This means that during any communication over the phone, its geo-coordinates reveal the location of the person that used this mobile. From a technical perspective the same meta data is produced by both, successful and failed calls. Thus, unsuccessful communication also provide value for the surveillance operations.

Locational Surveillance via Silent SMS

A surveillance tool used very intensely in these investigations by German LEAs is silent SMS (also referred to as "Ping"). Besides eavesdropping on telephony, silent SMS were used extensively to identify the person's location and whether the mobile phone was switched on or off.[13] What is a silent SMS? A silent SMS is sent to a mobile phone to obtain location data, the approximate whereabouts of the device. A silent SMS does not notify the receiver of its emergence and contains no content data; it is not perceivable to the user. But as it is an SMS in the technical sense: it generates meta data, which then provides insights into the whereabouts of the phone. Figure 8.2 explains one schedule of sending silent SMS and the LEAs reasons for doing so. The number of silent SMS sent during the

---

10 A code to uniquely identify a mobile phone operator/carrier using the one of the common mobile communications standards such as GSM.

11 The MSC is the primary service delivery node for GSM/CDMA, responsible for routing voice calls and SMS as well as other services. The MSC sets up and releases the end-to-end connection, handles mobility and hand-over requirements during the call and takes care of charging and real time prepaid account monitoring.

12 IMSI is the International Mobile Subscriber Identity, a unique number which identifies a SIM card. SIM cards often link to the user's name.

13 The latter usually lead the LEAs to the assumption that if the accused switched her phone off, she is suspicious of preparing or committing a crime.

Vorgangsnummer: V070015370 Seite: 1

T-Mobile
Auskunftsersuchen
Verbindungsdaten Service-Provider-Kunden

Verbindungsdaten vom: 01.10.2006 00:00 bis: 31.03.2007 23:59

Suche nach: MSISDN: +49(171)[redacted]

Auftragsart: nicht zyklisch
Service Provider: debitel AG (XTRA)

1. Auskünfte über Call-Records

Geprächsbeginn	Menge	CTT	Dienst	Type	B-Teilnehmer-RufNr	Zuatzdienste	MCC	MNC	MSC-ID	Cell ID	Länge/Breite/Richt. [1]
03.10.2006 19:31:13	3	2	B26	0	+49 171 [redacted]		262	01		0	
04.10.2006 13:38:10	17	2		2	30[redacted]	CLIP	262	01	03070000	42274	
05.10.2006 16:01:27	52	2		2	30[redacted]	CLIP	262	01	03070000	62933	131803/523019/120
05.10.2006 16:09:52	144	2	T11	0	+49 171 [redacted]		262	01		0	
05.10.2006 16:12:48	29	1	T11	0	33[redacted]		262	01		32974	131550/522918/240
07.10.2006 11:54:27	14	2		2	338[redacted]		262	01	03060000	13496	
11.10.2006 13:14:15	193	2		2	338[redacted]		262	01	03070000	42274	
12.10.2006 13:18:07	0	16	T22	0	+49 179[redacted]		262	01		15275	
12.10.2006 13:19:44	0	16	T22	0	+49 151[redacted]		262	01		15275	
12.10.2006 13:20:03	0	16	T22	0	+49 179[redacted]		262	01		15275	
12.10.2006 13:43:03	3	2	B26	0	+49 171[redacted]		262	01		0	
12.10.2006 16:04:09	2	1	T11	0	+49 30[redacted]		262	01		15275	
12.10.2006 20:34:12	401	2		2	30[redacted]		262	01	03070000	41340	
13.10.2006 12:05:25	75	2		2	30[redacted]	CLIP	262	01	03070000	42274	
13.10.2006 17:22:40	14	1	T11	0	+49 179[redacted]		262	01		54654	
13.10.2006 17:55:54	50	1	T11	0	+49 151[redacted]		262	01		22671	
13.10.2006 17:57:28	27	1	T11	0	+49 179[redacted]		262	01		22671	
13.10.2006 18:26:33	30	2		2	151[redacted]	CLIP	262	01	03060000	22671	
13.10.2006 18:46:25	18	2		2	179[redacted]	CLIP	262	01	03060000	22669	
13.10.2006 19:50:09	58	2		2	179[redacted]	CLIP	262	01	03060000	22669	
16.10.2006 14:05:48	161	2		2	338[redacted]		262	01	03070000	38271	
18.10.2006 19:04:34	35	2		2	30[redacted]		262	01	03070000	45131	132026/522743/0
19.10.2006 21:45:56	67	2		2	179[redacted]	CLIP	262	01	03070000	21996	

Figure 8.1: Example of meta data from mobile telephony retained by a telecommunications provider

Dem Beschuldigten [redacted] werden zu festgelegten Uhrzeiten *(10:00 Uhr, 20:00 Uhr, 00:00 Uhr – 04:00 Uhr)* sog. „Pings" (stille sms) auf sein Mobiltelefon gesandt, um feststellen zu können, ob sein Telefon – wie in verschiedenen linken Szenezeitschriften allgemein während konspirativer Treffen und der Begehung von Straftaten empfohlen – ausgeschaltet wird und sich somit Hinweise auf konspiratives Verhalten ergeben.
Zudem ermöglicht die Zustellung eines "Pings" die Bestimmung der Funkzelle, in der das Handy eingebucht ist.

Bei Observationen wurden zusätzlich anlassbezogen „Pings" versandt.

Figure 8.2: The practice of sending silent SMS: "On predefined times (10:00h, 20:00h, 00:00h - 4:00h) so called "Pings" (silent SMS) are being sent to the accused mobile phone to evaluate if his phone – as recommended in various leftist scene publication generally during conspiratorial meetings and during committing a crime – is switched off and thus hints towards conspiratorial behavior are given. Additionally, the delivery of a "ping" allows for identification of the cell, which the phone is logged in. During observations, additional event-related "pings" have been sent" [Translation O.L.]).

whole investigation is in the range of tens of thousands. They were all sent by the authorities themselves to generate connection data, which the LEAs then requested in the form of an information request from the telecommunication providers. The delay between request and reception of connection data varied between one week and one month. Thus, silent SMS is a measure for long term surveillance.

Figure 8.3 is an evaluation of hourly sent silent SMS to one mobile phone. Since some PINGs were not succesful, suspicion grew, although technical issues might have played a significant role here.

IMSI and IMEI

A common practice among activists is to change the SIM Card to obtain a new number and thus a new profile. Figure 8.4 and figure 8.5 show that both the SIM card related International Mobile Subscriber Identity (IMSI) and the device related

Die Auswertung der seit dem 16.11.2006 stündlich - mit Ausnahme von 2 Unterbrechungen aufgrund technischer Probleme – gesendeten PINGs auf das Mobiltelefon des [redacted] ergab, dass sein Telefon in der Regel durchgängig empfangsbereit war. Allerdings wurden auch Zeiträume festgestellt in denen die ausgesendeten PINGs nicht ankamen, was darauf deutet, dass das Handy ausgeschaltet gewesen sein könnte. Teilweise ließen sich über die Überwachungsmaßnahmen Erklärungen für dieses Verhalten finden (s.o.). Andere Zeiträume der Nicht-Zustellung von PINGs fanden bis heute keine Erklärung. So wurden am 20.11.2006 ab 3:00 Uhr keine PINGs mehr zugestellt. Erst um 08:16 Uhr erfolgte wieder eine Auslieferung. Warum das Mobiltelefon in dieser Zeit vermutlich ausgeschaltet war, konnt bislang noch nicht ermittelt werden.

Figure 8.3: Evaluation of silent SMS sent to one mobile phone. "The evaluation of the hourly sent (since 16.11.2006) PING's to the mobile phone of XX showed that his phone had been able to receive SMS generally. Nonetheless there have been time frames where the sent PING's did not arrive, which might mean that the phone has been switched off. In part this can be explained by other surveillance measures taken. Other unsuccessful deliveries of PING's cannot be explained until today. As on the 20.11.2006 from 3:00 AM onwards until 08:16 AM no PING's could be delivered. Until now it has not been clarified why the mobile phone has been switched off during this time" [Translation O.L.].

NL 2 30 TelH78 öKSi 453K052
Rufnummer: 030 / 4540052
Standort: Tegeler Str. 38, 13353 Berlin

die Mobilfunknummer des [redacted] angerufen wurde. Die Verbindung bestand 36 Sekunden, bei Zustandekommen der Verbindung befand sich das Mobilfunktelefon in der Funkzelle mit der Cell-ID 65469 und den Geokoordinaten 52'29'45 N / 13'25'42 O. Für das angewählte Handy wurden zudem die Nummer der SIM-Karte (IMSI), sowie die Gerätenummer (IMEI) verzeichnet.

IMSI: 2620[redacted]
IMEI: 3576[redacted]

Figure 8.4: A precise description of one single call, referencing the address of the public phone booth that was used, duration of call, cell ID the mobile was logged into, the geo-coordinates of this cell and the mobile phone's IMSI and IMEI. "[From this public phone] the mobile number of XX has been called. The call time was 36 seconds, when the connection started the mobile phone was logged into the cell with the Cell-ID 65469 at the geo-coordinates 52'29'45 N / 13'25'42 O. IMSI and IMEI from the called mobile phone were recorded." [Translation O.L.]

International Mobile (Station) Equipment Identity (IMEI)[14] are transmitted in the network and thus used by the authorities. The Location Area Cell (LAC) denotes the whereabouts of the phone in a numerical code.

> `Changing sim cards doesn't really do anything.`
>
> Freitas / NYC

Nathan Freitas' skeptical view on changing SIM cards as a means to remain anonymous is confirmed in that the IMEI, which is hard coded into the device, is harvested by the authorities as well as the IMSI.

14 IMEI is the International Mobile Station Equipment Identity, a 15-digits long, unique serial number, which makes it possible to identify any GSM or UMTS client.

	Beginn	Menge	CTT	Anrufende MSISDN	Angerufene MSISDN	Angerufene IMSI	Angerufene IMEI	Angerufene LAC	Länge / Breite
Call	24.07.2007, 03:35:20	36	22	'4930[redacted]'	'49160[redacted]'	'2620[redacted]'	'35769[redacted]'	65469	132542/522945
GSM	16.03.2007 02:38	21	2	30[redacted]				65469	132542/522945

Auszug aus Liste mit Verbindungsdaten T-Mobile zur Rufnummer 0160/[redacted]

Figure 8.5: IMSI and IMEI are part of the data sets provided by the telco.

Evaluation of one Specific Call Data

Figure 8.6 evaluates one specific call, generated on a special day, on which a militant activist action in Berlin took place. The officer evaluated (via Google Maps) that the average travel time from the crime site to the public phone booth, from where the surveilled mobile phone was called, allows the possibility that the person using the public phone also committed the crime.

Email Interception

As the accused subscribed to numerous of email lists, hundreds to thousands of emails went straight to the authorities during 2001 and 2008. This is an easy task for LEA, given the uncomplicated matter of getting access to email accounts of service providers such as GMX or Yahoo!. Interestingly, as figure 8.7 reveals, PGP encrypted emails could not be read. The email's subject of course is plain text, as well as its meta data. It states: "Four more PGP-encrypted emails have been found, whose subjects refer to the event in Trier. Nothing can be said about the contents of these e-mails."

Cross-referencing of Eavesdropping and Cell Logs

Figure 8.8 shows how both the results of eavesdropping and the data of a cell where a mobile phone was logged into are combined to evaluate one situation on one particular evening. Figure 8.9 states the wide knowledge the surveilling parties gathered about the suspects' social network, their activities and what the sister of one suspect said. Switching off one's mobile phone is interpreted as conspiratorial behavior.

> Bei der Auswertung der Verbindungsdaten des von [redacted] genutzten und auf seinen Namen angemeldeten Mobiltelefons mit der Rufnummer 0160 / [redacted] (T-Mobile) wurde festgestellt, dass in der Tatnacht um 02:38 Uhr[1] von der Telefonzelle
>
> **NL 2 30 TelH78 öKSi 453K052**
> **Rufnummer: 030 / [redacted]**
> **Standort: Tegeler Str. 38, 13353 Berlin**
>
> die Mobilfunknummer des [redacted] angerufen wurde. Die Verbindung bestand 21 Sekunden, bei Zustandekommen der Verbindung befand sich das Mobilfunktelefon in der Funkzelle mit der Cell-ID 65469 und den Geokoordinaten 52'29'45 N / 13'25'42 O.
>
> Laut Google Maps benötigt man für die 6,7 Kilometer lange Strecke vom Tatort am Märkischen Ufer bis zur Telefonzelle in der Tegeler Str. 38 zwölf Minuten.[2] Zur fraglichen (Nacht)-Zeit dürfte die benötigte Fahrzeit unter den angegebenen zwölf Minuten liegen.

Figure 8.6: "During the evaluation of connection data of the mobile phone used by XX and registered under his name with the call number 0160 / XXXXXX (T-Mobile) it was discovered that during the night of the crime at 02:38h [from this public phone] the mobile number of XX has been called. The call time was 21 seconds, when the connection started the mobile phone was logged into the cell with the Cell-ID 65469 at the geo-coordinates 52'29'45 N / 13'25'42 O. According to Google Maps it takes twelve minutes to travel the 6.7 km distance from the crime scene at Märkisches Ufer to the public phone at Tegeler Strasse 38. At night time the travel time should be under the specified twelve minutes." [Translation O.L.]

> Es wurden vier weitere PGP-verschlüsselte E-Mails festgestellt, die sich vom Betreff her mit der Veranstaltung in Trier beschäftigen. Zum Inhalt dieser Mails kann keine Aussage getroffen werden.

Figure 8.7: PGP-encrypted emails are not readable for the BKA.

Aus TKÜ-Maßnahmen ist bekannt, dass [redacted] ebenfalls in die Bar „Alois S." zu [redacted] und [redacted] kommen wollte. Das teilt sie [redacted] in einem kurzen Gespräch um 21:03 mit. Lt. Funkzellenauswertung des Handy von [redacted] war das Handy um 23:00 Uhr in der Funkzeile Lilli-Henoch-Str. und um 24:00 Uhr in der Funkzelle Lettestr. eingebucht gewesen ist.
Es steht daher zu vermuten, dass [redacted] und [redacted] die Bar gemeinsam verlassen haben und sich zur Wohnung des [redacted] begeben haben.

Figure 8.8: Reference to a combination of eavesdropping on telephone calls and cell log data (obtained through silent SMS most likely). "From eavesdropping operations it is known that XX as well wanted to come to the pub Alois S. to meet with XX and XX. This she told XX in a short phone call at 21:03h. According to the evaluation of cells the mobile phone of XX was logged in, it was logged into the cell Lillie-Henoch-Street at 23.00h and Lettestreet at 24:00h. Therefore it is reasonable that XX and XX left the bar together and went to the apartment of XX." [Translation O.L.]

The document is actually written to get permission for a prolongation of the surveillance operations from a judge. With only very few modifications it has been used over and over again to get these permissions from the same judge over and over again. It references a combination of eavesdropping on telephone calls and cell log data (obtained through silent SMS most likely).

Conclusions from the Documents

> People don't think about the infrastructure and so dataveillance' threat is a threat that you can't even necessarily imagine unless you spent a decent amount of time to figure out how the technology actually functions
>
> Kahn Gillmor / NYC

All the example documents given above regarding the surveillance operations on the suspects point to the infrastructure as the source of interception. The infrastructure offers a point of interception without much risk of detection. When communication was encrypted client to client, its contents were secured against this surveillance scheme, as figure 8.7 showed.

Durch die TKÜ-Maßnahme konnten umfangreiche Informationen über die Kontakte und das Engagement des Beschuldigten in verschiedenen Zusammenhängen gewonnen werden. Insbesondere konnten die Kontakte der Beschuldigten untereinander über Telefon und per E-Mail belegt werden und mehrfach die Verabredung persönlicher Treffen festgestellt werden. Auch wenn die Inhalte der TKÜ im Überwachungszeitraum keinen Aufschluss über eine Tatbeteiligung des Beschuldigten gaben, so erbrachten die Erkenntnisse aus der TKÜ in Verbindung mit der Zustellung von "Pings" auf die Handys der Beschuldigten [redacted], [redacted] und [redacted] Hinweise auf ein konspiratives Verhalten (Ausschalten der Handys) anlässlich eines Treffens zwischen [redacted] und [redacted] am 22.11.06. Darüber hinaus ist aufgrund der scherzhaften Bemerkung der Schwester des [redacted] nicht auszuschließen, dass sie entweder Kenntnis von der Verübung entsprechender Straftaten ihres Bruders hat oder dies zumindest für möglich hält.

Figure 8.9: "Extensive information could be gathered by eavesdropping of telephone about the contacts and the engagement of the accused in different interrelationships. Specifically the contacts of the accused amongst themselves via telephone and email could be proven and multiple personal meetings have been identified. Even if the contents of the eavesdropping during the time of surveillance did not provide details about a participation of the accused in the crimes, still, the results from the eavesdropping operations in relation with the transmission of "Pings" to the mobile phone of the accused XX and XX showed conspiratorial behavior (switching off the mobile phone) on the occasion of a meeting of XX and XX on the 22.11.06. Additionally, due to the joke of the sister of XX, it cannot be excluded that she either knows about his brother committing such crimes or at least thinks it is possible." [Translation O.L.]

That location information was actively generated by sending a very high number of silent SMS over a long duration, additionally points to the strategic importance of the cellular network for LEAs. The location data on the whereabouts of the suspects provided by the suspect's mobile phone are sufficiently precise for the LEAs. In a city like Berlin mobile phone cells are densely rolled out. The location data which is as exact as the functional radius of the cell is quite precise. The location data provided with silent SMS and cell log data generates a history of the suspect's travel route and whereabouts on an hourly frequency. Once the mobile phone had been switched off (or the battery was empty), this triggered an alert for the authorities and these time frames then were considered as a conspiratorial behavior.

Switching off the phone actually is a common practice of activists to defend themselves against surveillance.

> Regarding mobile phone, if we have any action plan, and if we know the issue is sensitive or the movement is in a sensitive moment, we would not explicitely speak about it, and at the meeting, we take out all the batteries of the mobile phones until the meeting is over.
>
> Freddie / Hong Kong

But this practice directly triggered suspicion, and the LEAs in the documented case argued that this practice signifies conspiratorial behavior, in order to get prolongations for the surveillance operations from the courts. The judge, apparently, agreed and allowed the authorities to continue with a total surveillance programme. But there are many reasons why a mobile phone does not work. Battery life is one. Or maybe the reason to switch it off is to have some privacy in the old fashioned sense: to not get disturbed, especially at night time.

Comparing security considerations of the activists, and the practice of the German LEA in the MG investigation, it becomes evident that no real measures, besides email encryption and switching off phones, are readily available in keeping mobile communications an effective and secure tool without risking the inherent support of surveillance operations. The strong accounts provided by meta data about the social networks of the suspects (who called whom, and when and where) can hardly be countered.

> I don't know about other countries, but in Japan the police is a very strong organisation. They collect everything they can.
>
> Yasuda / Tokyo

The nature of the relationship between resistance to surveillance and the answers of surveillance practice, to these kinds of resistance practices, can be understood as dialectical. Activists use code words, limit mobile phone use and even use encryption. The surveillance practice thus turns to meta data analysis and even produces them, as in the case of silent SMS. If resistance goes as far as to dismiss the use of mobile telephony altogether by switching the phones off, LEAs' surveillance practice answers by prolonging their surveillance activities with the argument that it is suspicious if activists switch off their phones (even though there is no reliable evidence of criminal activity).

The relationship of resistance and surveillance can be seen as circular or spiral: once one side re-empowers itself by technical measures, the other finds new ways to gather information or better hide their activities. There is a limit here, of course, which is the loss of agency: in a society that relies on telecommunication, activism cannot dismiss its use completely. Avoiding or minimizing the use of mobile phones can inhibit agency; similarly, too many security measures can hamper the effectiveness of activism.

Part Three: Bottom-Up Strategies

9 Necessities for Trusted Telecommunication

> Actually, the problem here is mainly the killing of activists and journalists. The son of a press freedom icon, the leader of the alternative press during the Marcos years, his son was abducted two or three years ago and not yet released. It is suspected that he died
>
> Eder / Manila

Maybe the danger of killing activists can be reduced if appropriate measures are taken in the way how activists communicate. The technologies of communication for this purpose need to be hardened, so that the many possible threats can be made at least made more expensive. This chapter thus explores the possibilities at hands to secure communication against eavesdropping, and dataveillance in general. I aim to show that it has become increasingly hard to remain under the radar as surveillance technologies have caught up to securing measures. Additionally, in the last years policies have been released which deliberately attack trustful communications. These neo-sovereign measures, as mentioned in chapter 7, belong to a panspectron which surveils, collects and mines all kinds of data. Targeted dataveillance has also increased and become a more and more common tool in the arsenal of tools of LEAs, as chapter 8 has shown. The use of Trojan software, which agencies install on target hosts, and which report activities and copy the content of data storage devices, is another prominent case.[1]

---

1 Euphemistically called 'remote forensics', the intrusion into private computers of surveillance targets by German LEAs has been widely criticized when the Chaos Communication Club investigated the software. Besides the general question if such means are lawful, it is unlikely that information gathered by state Trojan software can be used in courts at all, as it alters the contents of the hard drives itself. Thus, it can only be used for spying. See https://www.ccc.de/en/updates/2011/staatstrojaner (accessed 2 March 2012).

In a book already published in 1998, a time when email and the web started to take off very quickly, but were still in a nascent state, two well-known scholars of cryptography, Susan Landau and Whitfield Diffie[2] write that:

> The potential impact on privacy is profound. Telecommunications are intrinsically interceptable, and this interceptability has by and large been enhanced by digital technology. Communications designed to be sorted and switched by digital computers can be sorted and recorded by digital computers. Common-channeled signaling, broadcast networks, and communication satellites facilitate interception on a grand scale previously unknown. Laws will not change these facts. (Diffie and Landau 1998, 226)

If "laws will not change these facts", maybe technology itself could. Therefore, a conceptual and in part technical discussion on how electronically mediated communications can be hardened, i.e. made robust and more trustworthy to their users, is necessary to investigate this.

Chapter 8 has shown that when privacy rights are outperformed by investigation means, mobile media turns into surveillance media; it changes its modality. Both in terms of contents, by machinic or human eavesdropping on the calls and SMS, but also as a location tracker, which reports the whereabouts of the target. The fact that agencies go as far as to communicate with the mobile media device themselves, by sending silent SMS, leads to the assumption that the device is actively used by two parties, the target and the surveilling party; dual use, one might say. Like Trojan software, silent SMS go beyond a passive sniffing: they actively intervene into the devices' activities. Both thus cross a boundary, which on the one hand makes

2 Diffie and his colleague Hellman were the first to propose an encryption algorithm where the encryption and decryption keys are different, so that one can publish the encryption key and keep the decryption key private. "Diffie and Hellman's idea is based on complexity: some mathematical problems are easy to compute but their inverse appears computationally difficult. Integer multiplication and the inverse problem, factoring integers into their prime factors, seem to be one such pair" (Landau 2010, 45). However, Diffie is most prominent for the actual realization to establish an encrypted connection without the need for the parties involved to have shared a secret beforehand. "Anyone can now send you an encrypted message, and only you can decrypt it. [...] The DH [Diffie-Hellmann, O.L.] protocol is a real nifty idea. It turns out that two people communicating over an insecure line can agree on a secret key in such a way that both of them receive the same key without divulging it to someone who is listening in on their conversation." (Ferguson and Schneier 2003, 208-209)

them more prone to discovery. By using the company owned infrastructure for *sending* SMS, the LEAs are becoming customers. Billing and all other business procedures apply. The heterogeneous ensemble of telecom providers, a chain of actors and machines, become part of the process.

On the other hand, both mark a tipping point of liberal rule, that signifies that the activities of the agencies are taking advantage of a technology of governmentality, while they do not interfere with the governmental scheme of mobile media.

In general, all interviewees expressed their expectations about surveillance clearly in that basically they would not expect privacy a strong part of their communication rights anymore.

> I think that at a given moment we can be the subjects of surveillance. Police has the interest and the resources to practice surveillance. And there has been recently the creation of the cybernetic police on the federal level. There is a similar entity also in Mexico City. It is also well known there are private technical teams, hired as mercinaries, to do these jobs. So, there is also a risk for us to be under surveillance.
>
> Enrique / Mexico City

As there are no machines of liberation, it is decisive not to fall into the trap to think that secure, trustworthy tele-communications means provide a solution for many things. This I have tried to show in the previous chapters: that communication technologies are embedded in a larger set of governmental practices and different modalities of rule. In this chapter, I do not intend to fall behind these claims. Nonetheless, I argue, the diminishment of risk to exposure via surveillance means can still facilitate a step towards independence from neo-sovereignty programs; it can be a way to outperform state measures that use a governmental technology for sovereign purposes and, finally, reclaim mobile media in an autonomous, self-governed mode.

The first step then is to discuss software as a product of societal relations, that both reproduces patterns and trajectories of society, and can be seen as a sediment of such relations.

Software is Part of Society

This hints towards agency not yet discussed: that the use of technologies goes hand in hand with the way it is rolled out and designed. Most importantly, the process of its production, its modalities of production, cannot be excluded from the use it entails: where the GSM network is in general terms inaccessible to non-commercial entities, at the same time it provides the most easy point of surveillance. In contrast, the Internet, which is based on open standards and in large parts open source software and protocols, provides access to tool fabrication for non-commercial parties, such as the huge free/libre open source software community. This has led to several software solutions that provide privacy and trustworthiness to its users – of course within limits. Where source code is publicly accessible, it is open for debate, followed by updates and bug fixing. Transparency in the production of communication technologies can thus be seen as a powerful modality of technology production that encounters the commercial paradigm of closed source: non-disclosure agreements and many more obstacles make it impossible to review closed source software.[3] This does not necessarily mean that open source software is always 'more secure'. But the trajectory of transparency directly transports security as a matter for the public.

With respect to the power of code as a regulating instance, and execution power, Lessig (1999) points to this contested area and argues that code has become the regulating principle of societies that rely on networked computers. Code, he argues, can be seen as a new law, the law of cyberspace. This makes those that produce it, if one follows the analogy, nothing less than lawmakers. Grassmuck (2002) problematizes as well the order of law and intellectual property in regard to software and recounts how free and open software is mostly a collective process that subverts the analog regimes of artificial scarcity that are applied on commercial, closed source software. He argues that free and libre open source software, or FLOSS, is a puzzle piece for new societal relations which aim at an open and public repository of knowledge.

An activist I interviewed in Madrid, who is involved in open source software production and hacking, refers to the larger frame of software and technology in this way:

> There are always two sides, one is in relation to the external everyday life world, which is traversed by power structures, commercial agencies. It is to try to open some room to stop some trends on the

3 In Leistert (2012) I discuss the difference of mobile devices that allow open source software to run and give access to large parts of their operation system and those, which are sealed and closed, up to the point that taking out the battery voids the warranty.

> streets that are not fair or: we want social justice, to capture the whole idea. Social justice is also about interpersonal relationships, not only super powers. But in particular our goal is to work on the dimension of technology. Part of the success of the movements is also the technology in many aspects. When we communicate within an assembly, when we organize as a network, we always use technologies. Language is a super power technology. The space in the hacklab is technology as well, not only bits and keyboards. And when you start thinking that technology is power, both empowering power, not one that you use to oppress others, suddenly it gets a dimension that covers almost everything and be very effective if you are skillful. The intention would be to open free spaces where people can control their technology and be empowered. Once we have opened spaces, like free software, we also collaborate and produce, and in particular inside social movements, in a context that is already social and political in many other aspects, we contribute with the technology and the skills we have.
>
> Anonymous2 / Madrid

This shows that in groups, which are already sensitive towards power relations and provide practices of solidarity, free and open software is not merely seen as a gift, but as a political tool to change societal relations. This notion will return in the discussion of self-controlled infrastructure.

Software is a Cultural Artifact

A slightly different approach, one that files software in the register of a cultural continuum, reintegrating the reified software into a larger process, are *code studies*. The programatic lexicon of code studies, edited by Fuller (2008) lists some of the hegemonic concepts that are realized in software and discusses them from a cultural studies/humanities perspective as digital objects. Such concepts are 'algorithm', 'pixel', 'interaction', or 'function'. "Technologisation of the senses and structuring of relations by technology is often carried out by naturalized means; lessening our ability to understand and engage with such changes." (Fuller 2008, 4). Code or software studies can be understood as an approach to de-naturalize these means, and make it possible to engage with these changes.[4]

> It depends on how we appropriate technology. If we were using more the possibility to develop software that would facilitate the

4 On the relation of performativity, repetition, language and code, see Winkler (2004, 226-229).

> empowerment of communities and that could be very important. The use of free software and the appropriation of the technology would be responding to the need of the communities.
>
> Enrique / Mexico City

"Protocol" by Galloway (2004) argues in the same direction: Galloway focuses on the regime that controls the conditions of information exchange between layers and entities in a technical network. Less code studies, more digital architecture studies, he also engages in a discussion about naturalized digital environments. "Protocol is a technique for achieving voluntary regulation within a contingent environment." Galloway (2004, 7) And he specifies "protocol is a distributed management system that allows control to exits within a heterogeneous material milieu" Galloway (2004, 8). By tracing its genealogies, objects and actors, he differentiates a larger field of heterogeneous forces that struggle within the immanent regime of protocol management.[5]

The societal and cultural approaches towards software show that mobile media as a semiotic-material conglomerate enmeshed in a variety of governmental and sovereign programs relies in part on how it is produced as a technology in the stricter sense. This black-boxing is part of the power relations it executes. Software thus can be seen as a thickened, condensed, executable function that inherits many kinds of amalgamations from fields and areas that are not easily traceable unless one reads and writes code.

Social Engineering

> My concern is that the fights against surveillance in the abstract are so denatured of their political context that they might not be entirely convincing
>
> Kahn Gillmor / NYC

When this thickened powerful tool is in wide use, again cultural and societal aspects re-inscribe themselves into code, in the way of passwords, that are chosen, or in more general terms, what it is used for, what it facilitates. This 'human factor', (to look a it from a technocentric perspective) very often remains the weakest

5 I have discussed protocological power regimes in Leistert (2010), where I specifically refer to wireless network topologies as sediments of societal relations.

point in terms of security. The two classical reports about "controlling the human element of security" are the two books presented by former hacker Mitnick (2002; 2005). Mitnick was sentenced for the intrusion into computer systems of many different companies. In these books, he describes that his point of departure in most cases was social engineering: gathering information about, but most importantly directly from the persons who operated the machines or had access to them.

> In internal discipline we are also concerned about social engineering. So, we try to avoid gossiping, that may expose sensitive information to someone maybe listening. And in MTST we had trouble not with big police but with local city police. So, gossiping may have gone bad and local police may have made something with something they may have heard of. And the police used that to sue people in the occupation.
>
> Fernão / São Paulo

A lot of what is done with digital communication tools depends on factors that are external to the machines. This depends, very much, on how it is used and in what kind of situation. Thus, taking care in many non-technological ways is part of a 'good practice', as Hijuchi from Tokyo states:

> When I was abroad, when I cross the borders, surveillance is always a problem, because I cross borders for political activities. Before I cross the border, I always prepare many things, contact people. Because I might be stopped at the border, especially when I prepare things before crossing the border, it can be a problem. Maybe I go to China next month to contact activists there. So I have to care a lot about this. In the recent years surveillance of the so called civil society is getting stronger here. E.g. the surveillance of many communities.
>
> Hijuchi / Tokyo

Globalized Technologies

The remaining part of this chapter's discussion is about technical concepts that enhance different aspects of privacy. While it remains open how applicable these concepts are in concrete situations, this discussion aims at providing a glimpse into the the techno-theoretical limits of both; surveillance and security.[6]

6 Security in the sense here refers to the provision of communication technology that does not or only with high costs allow for interception, eavesdropping or other tapping measures.

As this chapter does not focus on the very specific uses and practices, it is possible to neglect the specificities of locations and universalize to some extent. I share the assumption that

> when a pattern of conduct (for example, the substantial enhancement of individual and collective autonomy by wireless communication capability) repeats itself in several studies in several contexts, we consider it plausible that the observation properly reflects the new realm of social practice. (Castells, Fernandez-Ardèvol, Qiu, and Sey 2007, 3)

This new social practice can be confronted with the technological flaws in which it is embedded. In a rather technocentric approach from the view of an engineer, Landau (2010) treats these problems in this way:

> It does not help to tell people to be secure. In order for their communications to be secure, security must be built into their communications systems. It must be ubiquitous, from the phone to the central office and from the transmission of a cell phone to its base station to the communications infrastructure itself. (Landau 2010, 199)

While the technological reality is far from these demands of a high profile security architect, it is nevertheless encouraging to look at the possibilities.

To enusre that this discussion does not take place in a vacuum, in the next section the general reasons why a secured telecommunication environment is important are reconsidered. The argument is that the enfolding of a political discourse benefits from a trusted, secured communication environment.

Re: What are Trusted Environments Good for?

A political discourse relies on certain levels of trust. Trust here stands for what is often called 'privacy'. I do not relate to this term because it is confusing in many ways: why, one could ask, is privacy so important for public matters? Trust seems to be a more promising candidate as it covers a whole set of parameters which together form a trustworthy communication environment.If we cannot trust that the things we say will be not be used against us, our ability to continue candidly saying what we think is weakened. An open discussion, namely one without fear, is the basis for political discourse. Statements like this from Manila shows how a lack of trust invokes modes of self censorship or cultural techniques of obfuscation:

> We adjust. We change phones, although it is very expensive. E.g. when there is a rally tomorrow, we say there is a festival tomorrow.
>
> Mina, Minerva, Joan, Julie / Manila

Trusted communication environments therefore, are precisely a foundation for democratic processes.[7]

Whitfield Diffie and Susan Landau write that "the foundations of democracy rests upon privacy" (Diffie and Landau 1998, 127) and they argue that:

> Privacy is essential to political discourse. The fact is not immediately obvious, because the most familiar political discourse is public. History records political speeches, broadsides, pamphlets, and manifestos, not the quiet conversations among those who wrote them. Without the opportunity to discuss politics in private, however, the finished positions that appear in public might never be formulated. (Diffie and Landau 1998, 127)

Further, they state that "change often begins most tentatively, and political discussion often starts in private. Journalists need to operate in private when cultivating sources. Attorneys cannot properly defend their clients if their communications are not privileged." (Diffie and Landau 1998, 149) Needless to say that activists, who are amongst those seeking change towards social justice, are in strong need of the same circumstances. The following sections look into the pieces that together can build a trustworthy mobile media environment.In general, the same is true for other digital networked infrastructures, devices, and services.

Trust in Telecommunication 1: Securing Content

The general situation in telecommunication is different to 'natural' communication as the intermediary seldom is present or perceivable. It is mostly like a ghost,

7 Something Jürgen Habermas' (1987) theory of communicative rationality does not reflect on, as it rests on the assumption that such an environment is given as it enables an active democracy and a critical public. But telemediated communications barely ensure these conditions. Maybe his theory is outdated.

whose presence is hidden. In regard to technical issues with telemediated communication in general, this 'ghost' is possibly situated at any position; from the microphone or keyboard of the device to the respective interfaces of the device(s) the message is transmitted to.

> There were a lot of issues, like viruses, that ghost net virus that was reveiled recently was targeted at tibetan offices, essentially it would take over your drivers for your camera and microphone apparently. I have the source. It looks like it could do it.
>
> Freitas / NYC

And most likely anywhere in between: in the infrastructure. In a retrospective overview on efforts of the last 10 years to secure digitally mediated communications, Danezis and Gürses (2010) diagnose that all efforts so far have failed on a large scale. "Within their wider contexts privacy friendly communications only support some specific privacy needs." (2010, 12)

Although no 'one size fits all' solution has been created yet, in digital communications, where privacy and trust cannot be established via social means, as in face-to-face situations, mathematics can step in and provide trust on the level of digitality. This is possible mainly due to two developments in the last 40 years: provision of non-patented and fee-free crypto algorithms, and a defeat of the mainly US long-term policy to forbid the unauthorized dissemination of cryptography. While it is true that "the centrality of mathematics to privacy in the electronic environment [...] creates new privacy classes" (Braman 2006a, 131), today's 'commoners' can rely on open source, non-patented tools. Plus, encryption's "use outside of government became legal in the United States only after many years of political struggle"(Braman 2006a, 131). While the attempts of different government bodies to regain control over encryption are numerous and still prevail, the scene is set and it is hard to imagine that a reversal is possible. "Widespread availability of cryptography was here to stay." (Landau 2010, 2) One of the best known cases of "encryption for the people" is Phil Zimmerman's Pretty Good Privacy (PGP) software, whose source code in its very early state was even printed and exported as a book to fall under the right of free speech and in this way the control of weapon export law was circumvented.[8] The best known and most widely used crypto-

8 MIT published PGP's C source code as a book. The still informative preface by Phil Zimmerman is on-line: http://www.philzimmermann.com/EN/essays/BookPreface.html (accessed 4 March 2012)

applications (OpenSSL/GnuTLS) and OpenPGP tools are available open source, at no cost and without patent infringements.

Four Criteria for Trusted Communication

From a purely logical point of view, there are four steps for a digitally transmitted message sent from A to B that have to mastered: the message must not be altered, one must be able to know who has sent it, it must not be read by third parties and, the originator of the message must be able to deny having sent the message (this last point is not very intuitive, and I will come back to it).[9] This translates into the security concepts of

- Integrity
- Authentication
- Confidentiality ('Privacy')
- Deniability

Whatever happens afterwards with this message, even in the case it becomes public, does not alter these needs: these parameters are ways to empower the user to decide herself about the further processing of the message. By application of these measures, it becomes possible for users to set up a signaling path within an infrastructure controlled by adversaries. They provide in totalitarian all that is needed to communicate safely.[10]

Integrity

Integrity of a message is reached mathematically by generating a signature, which provides a link between the output of a mathematical calculation done by the originator of the message (typically referred to as Alice) and the recipient (typically

9 Anonymity is a special case that I leave out here. It is important socially from a security/safety perspective for one party to be able to anonymously create a confidential, integrity-checked communication with a well-authenticated peer, for example the 'whistleblower' scenario.

10 This section here is about the content. The issue of the *fact* that communication between A and B takes place, and the powers to surveil this, are discussed in 9.

referred to as Bob), who in the same way can calculate the signature and if the content of a message was altered. In short, these *Message Authentication Codes* are values that can be computed by both sides of the message. The attacker, usually referred to as Eve, "cannot alter the message without Bob detecting it." (Housley and Polk 2001, 9)

The lack of such mechanism allows that Eve may intercept a SMS, cut out the middle part of the contents (or replace it with something else), play it back, so that Bob has no immediate way to find out about it. The same occurs with emails: no mechanism ensures the integrity of an email. TCP/IP data transmission is not safeguarded against this. Neither is Simple Mail Transfer Protocol (SMTP).

Prividing as no additional mechanism is implemented to perform on the integrity of a message, the integrity cannot be guaranteed. Telecommunication infrastructure are forge-prone once an intermediary is settled between the sending and receiving device. On the level of the established protocols alone there is, unfortunately by default, no such safeguard.

With the application of an integrity check of a message, the content of the message is secured against alteration of its contents.

Authentication

Authentication of a message in telecommunication is crucial, because as a technology of distance, all means that are usually available to authenticate someone are not at hand. This starts with telephone calls if the caller's voice is unknown.

Even more problematic is written communication, such as emails. Maybe the style and use of some words is familiar in an email, but not at all, or enough to be sure that the email was written by the person claiming to have done so. One of the easiest hacking tasks is to spoof email addresses, a phenomena very well known due to email spam, which often pretends to have been sent by a friend or colleague by simply displaying their email-addresses in the `from-field`, while the `return-field` differs and leads somewhere else. The protocol to send email, SMTP, allows unauthenticated sending of emails, although more and more providers are taking measures to restrict this.

SMS origins are also very easy technically alterable, as there is no dedicated mechanism of authentication built into the SMS protocol. To give one example how the SMS protocol allows the insertion of an alternative reply address.

> "[D]uring message submission, the message originator has the capability to indicate that message replies should be sent to an alternate reply address. This is performed by inserting, in the submitted message, a dedicated information element containing the alternate reply address." (Le Bodic 2005, 107)

But how can a message be trusted if it is not clear who has sent it? The possible consequences might be tragic. Regarding the case of activism, a message sent from one manifestation to another saying that everything was alright, while in fact the demonstration was halted by the police, may actually harm the whole planned action. Daniel Kahn-Gillmor describes his worries:

> Another one of my concerns and this has to do with data autonomy and communications autonomy is also the lack of authentication in a lot of these things. And there is enough concern about infiltration by either law enforcement provocateurs or by direct antagonists. Someone is sending a message, saying "come on over here and rescue your buddy" is like: what does it mean in a situation where you are not under pressure and what does it mean when you are under pressure and time is the essence? How does that work out? In many situations that I have seen there has not been adaequate groundwork laid and this is coming from an engineering standpoint. As somebody who has training as a software developer. And a lot of the systems that I have seen I thought "oh oh"
>
> Kahn Gillmor / NYC

As with face-to-face communication, where the authentication of message and mouth, so to say, is unproblematic, we intuitively expect the same for telecommunication. But if it is not possible to authenticate a message, it is unauthenticated and strictly speaking cannot be trusted.

This missing reference somehow has to be reintroduced into telecommunicated messages, to regain authentication. Again, this can be done mathematically, via the digital signature, which is calculated from Alice's private key. As Bob possesses Alice's public key, he can mathematically prove that the message has been signed by Alice, thus proving the authenticity of the message.[11]

---

11 There are in fact many more means to authenticate a user against a system or the other way round, starting with password authentication, one-time authentication values, or even such sophisticated systems as time-based implicit challenges. But for the discussion here, they are not relevant.

Confidentiality

> If we can encrypt the messages we are sending, this would be very useful.
>
> Mina, Minerva, Joan, Julie / Manila

Confidentiality is the most intuitive part, often confused with privacy. But all four points listed here matter for privacy. Confidentiality has only a single goal: to provide the means, that Eve cannot read the messages that Alice sends to Bob. In a hostile communication environment, that is controlled by intermediaries (Eve), any point in the infrastructure allows the opportunity to copy the communication and analyze it. This can be done at various places, wherever the message passes. Encryption of messages is thus the technical answer to message eavesdropping.[12] Many activists interviewed have expressed their familiarity with encryption, although nearly none in the field of mobile media. Encryption of messages remains a domain of email and chat.

> Sometime ago, a collegue in San Cristobal asked to receive email only encrypted. He said: even if you only say "Hello". It is a period for 2 or 3 years now, that we try to encrypt emails. Between 1995 and 1998 we made workshops about that, but by now people have forgotten about it
>
> Olinca / Mexico City

There are very different methods for encryption, and its sub-necessities like key generation, session-handling and so forth. This field is not be covered here. For a very good overview on the mathematics involved, see Ferguson and Schneier (2003), for how to apply this, see Schneier (1996) and Menezes, Oorschot, Vanstone, and Rivest (1997). A good overview on actual software that enhances confidentiality provide Acquisti, De Capitani di Vimercati, Gritzalis, and Lambrinoudakis (2007).

> We use GPG for email. We use TLS for the website.
>
> Yasuda / Tokyo

12 Encrypting communication in fact is an everyday operation performed by billions of online customers to shops or banks that use Hypertext Transfer Protocol Secure (HTTPS). But it must be stated that more and more people in the IT security community regard the certificate authority (CA) structure, which is used to authenticate the certificates for the users, as heavily compromised. See Soghoian and Stamm (2011). Until now, there is no viable solution to this severe problem.

Often, an analogy is used to illustrate what encryption provides for confidentiality: to compare unencrypted emails with postcards, which can be read by everyone that transmits/transports it, and encrypted email with letters, which are sealed in an envelope and thus not everyone can read.[13]

Iokese, a tech-savvy activist, lists the technologies of his practice to enhance privacy of communication:

> Encryption: in email TLS/SSL, in Jabber OTR and SSL, and VPN to access data. But I do not encrypt my emails by default, because 90% don't use email encryption.
>
> Iokese / Madrid

Deniability

A rather counter-intuitive problem is deniability. It addresses the impossibility to rescan the past acts of communication. Deniability offers the technological possibility to deny having said this or that. Because even if the communication is encrypted, deniability is not always given. Only a few encryption schemes actually offer this feature by using a new encryption key, or new elements that seed the encryption for each message. This is referred to as perfect forward secrecy (PFS). It is most prominently a feature of Off-the-Record Messaging (OTR).

If Alice has already sent Bob a hundred messages encrypted with Bob's public key, and Bob's private key falls into the hands of Eve, then Eve can decrypt all hundred messages in retrospect, in the case that PFS is not part of the encryption design. PFS technically ensures that if one message is decrypted by Eve, the past and future ones are not affected. Thus, they can be denied by the sender. It has a paralell to the transience of the spoken word.

Deniability in very strict terms does not count amongst the core principle's of a trusted digital communication. But to my understanding, this has historical reasons, not so much logical ones, as it only became feasible in recent years. See Borisov and Goldberg (2005) and Engel (2007) for details.

---

13 This metaphor has a problem though: while an envelope can be opened without notice by trained personnel, forced decryption of emails, if encrypted properly, is very machine intense and thus too expensive for LEAs in general.

Example SMS: Simply Monitor Someone

> There is a specific directive to the national telecommunications office. They store all messages for 3 months and then they garbage it, but recently they extended that to 6 months, or up to one year. They keep it longer now. So they can go back to anything and track it down.
>
> Mina, Minerva, Joan, Julie / Manila

To summarize: to become a trustworthy channel for activists, SMS, the service used commonly by most of the interviewees' groups, would need: authenticity to guarantee that the receiving party can be sure about the authenticity of the sender; integrity to ensure that the message has not been altered; confidentiality to make sure that no one else except the intended partakers can read the contents. And finally, deniability would ensure that even if the message has been compromised in some way, this does not affect past or future SMS communication.

> I try not to use my mobile for critical contents, I tend to use for that a landline. I prefer to meet people. Once we organized 5000 people: we got a call from the communities saying there is a public hearing and we need your help to mobilise and they gave us two days time. We arrived one day before the public hearing and we met with the key leaders. So we thought this is simple, we just phone them. But they said: no don't do that, turn off your phone, I want you to go house to house, mobiliZe them. It worked, next day there were 4000 people. If we had done it through SMS, there would have been counter-mobilization. So tactically, it was useful not to use it. It is so much easier for the police to tap you than to go and stand and watch you.
>
> Saldanha / Bangalore

Unfortunately, none of these criteria is met by SMS: it does not provide any such mechanisms. There is a mild confusion about whether SMS transmission is actually encrypted. Susan Landau states such a thing:

> SMS messages are essentially broadcasts between the cell tower and a cell phone. As such they are easy to intercept. The messages are encrypted. However, the type of encryption, which varies with the type of network, is not considered very secure. (Landau 2010, 178)

It is important to state that only the last node, between cell tower and cell phone, might be weakly encrypted. Additionally, this is not a necessity to run a cell, which leads to many cells that do not ask for encryption for the connection to the phone.[14]

The other general issue with SMS, and in general with GSM, is a real tough problem: there is no authentication.

> The bigger security gap in GSM, however, was a known flaw that no one had expected to be much of a problem. To simplify the authentication of the user to the system, the GSM protocol did not require the cell tower to authenticate itself to the cell phone. This allowed fake towers to be set up and spoof phones in the network. (Landau 2010, 113)

Due to this flaw, IMSI catchers could be developed. These are fake cells, pretending to be the nearest cell to a mobile phone, and catching the phone by letting the phone log in. Then they proxy the traffic to the next real cell, while listening in on conversations, tracking the phone, or storing all data flows. As the name suggests, they are essentially useful to surveil specific IMSI numbers, i.e. only specific cell phones of suspects. For example, in the UK, IMSI Catchers go under the name *Datong Surveillance System*:

> The surveillance system has been procured by the Metropolitan police from Leeds-based company Datong plc, which counts the US Secret Service, the Ministry of Defence and regimes in the Middle East among its customers. Strictly classified under government protocol as 'Listed X', it can emit a signal over an area of up to an estimated 10 sq km, forcing hundreds of mobile phones per minute to release their unique IMSI and IMEI identity codes, which can be used to track a person's movements in real time.[15]

14 "Hackers crack open mobile network", BBC News, 31 December 2010. http://www.bbc.co.uk/news/technology-12094227 (accessed 21 March 2012).

15 "Met police using surveillance system to monitor mobile phones", Ryan Gallagher and Rajeev Syal, Guardian.co.uk, 30 October 2011, http://www.guardian.co.uk/uk/2011/oct/30/metropolitan-police-mobile-phone-surveillance (accessed 20 March 2012).

Integrity of a SMS cannot be provided for either. No mechanisms are built-in to check if the message has been altered. Nathan Freitas echoes that in terms of content analysis, 'authoritarian' regimes have invested in scanning technologies:

> The real test is when you are in Iran, Burma or China. Basically it is either SMS kind of level, or IP level, data packet inspection, keyword watching on packets, or voice analysis with complexity increasing
>
> Freitas / NYC

To return to the infrastructural flaws, Susan Landau sums up what is left if the architecture must be considered insecure:

> In a system that is itself architected insecurely – for example, with the potential for packet sniffers at routers throughout the infrastructure – two actions can substantially improve communication security. End-to-end encryption will ensure security of the message content. (Landau 2010, 199)

I will discuss the second "action", securing the transaction data, in the next section.

End-to-end encryption, if properly designed, can surpass the deficits of the infrastructure, because it renders the communication path autonomous in terms of security: a device to device privacy enhancing technology has the huge advantage of being independent from all infrastructural constraints and thus independent from the political regimes and their ability to access the infrastructure.

Sticking to the example of SMS: a client to client privacy enhancement software[16] has to be installed on each device independently.

> If someone is using the Internet to communicate, their communications security is entirely dependent on the application's security. Unless email and IM are encrypted or secured through a VPN, these communications can be eavesdropped on. (Landau 2010, 179)

The downside of all these approaches is plain to see: as it needs an extra effort, some extra skills, the user base for those tools is small. I agree that the "[t]he public, while in principle wanting private communications, in practice appears willing

16 For example *TextSecure* for Android phones, or *CryptoSMS.org* for J2ME feature phones.

to make it private only if the system is simple to use, does not affect the communication by slowing them down or degrading quality, and cheap" (Landau 2010, 47). Privacy enhancing technologies are only successful if their use and installations becomes a cultural technique as such, and is not perceived as a tedious, or at worst a frustrating task. But for smaller special interest groups, like activists, journalists, or human right advocates, privacy enhancing end-to-end solutions are feasible and applicable, taking their risk of exposure into consideration. This statement is backed by all interviewees: the general necessity to lay hands on one's own devices and install privacy enhancing software is seen as feasible.

Teletrust

> On the other hand, as long as the means of communicating is not controlled by the people who use it, I don't think there is much chance to get out of this problem that we constantly duck and hide, when we communicate and basically try to find workarounds to put security in a complete insecure model. Because constantly we rely on trusting what is happening in between or we don't give a shit. One of these two things.
>
> Startx / London

What Startx is referring to resonates within the findings of Danezis and Wittneben (2006), who have shown that while the communications between Alice and Bob might be secured against surveillance, this model is not realistic enough as it is highly probable that other members of their social network receive the same information. Thus one should "consider that information leaks through third parties" (2006, 13): "As it is often the case the innocent, who do not protect themselves, are the first victims of surveillance, while hardened targets remain elusive for a long time." (2006, 14). Therefore, the powers of mass surveillance as inhibited by data retention schemes, are only slowly recognized by security researchers. This leads to the final question: what about the infrastructure of communication? As Goldberg recalls: the massification of digital communications and especially its commercialization has changed practice and only a few services are left that are popular and used by everyone:

> One of the reasons I don't share the idea of only using free software is because the digital world doesn't look like it did five years ago. The Twitter example is one I am coming back to: at some point I decided that for most of my communication it is more important

> that we have insecure communication than that we have secured, but less accessful communication. I use GPG with everybody that has it, but even with people that were there at St. Paul and got handcuffed by the FBI, it is still hard to make them use it. I can not not communicate with them.
>
> Goldberg / NYC

This is seconded by a member of the Mexican infrastructure and service provider Laneta, which has been hosting services for activists without commercial interest since the 1990s:

> But now this has changed, there are now a lot of commercial, cheap services. The political aspect of ICT is lost with this a little bit. They have the feeling that they have all there needs resolved, but they lack the political point of view about the tools. A lot of people use it, but in terms of the commercial issue they are not thinking that this could be a political issue
>
> Olinca / Mexico City

Daniel Kahn Gillmor does not want to give up the idea of control over communications, a claim mainly tech-savvy interviewees shared:

> And then the other thing is the issue of control over communications. As a free software developer I do think (despite my caveats) that there is a potential for emancipatory change in a global communications network; but it can only happen when people are actually in control of their own communications. That is one of the reasons why free software is important. Mobile phones in the communications world have been our adversaries in this particular struggle. The idea of making myself dependent on a machine that has so many layers of inaccessible proprietary control is really unacceptable for me
>
> Kahn Gillmor / NYC

Some of the interviewees are engaged in setting up a parallel infrastructure which operates independently from policy and law, thus access for lawful interception is not provided. This is covered in the next section.

Trust in Telecommunication 2: Securing Transaction Data

It is very simple to put: transaction data cannot be secured in a way taht content data can be. Again Susan Landau states the point very nicely:

> As for security of transactional information, it is worth recalling NSA's thoughts on the matter: if the message were "really uninterceptable" then "the intended recipient, your own distant receiver, could not pick it up." In order for the message to be delivered, transactional information must be easily and widely available along the nodes of its route. (Landau 2010, 199)

> We moved to prepaid cellphones during the RNC and I did some research and found out that they actually have software that would find groups of phones that would only talk to each other on the premise that it is a criminal enterprise or what have you. And that is done automatically. That to me is dataveillance.
>
> Goldberg / NYC

In the strict sense then, a way to secure transaction data is not feasible.[17] But if the point of view is shifted slightly, a different perspective proves different: while in general terms this claim is true, it is not when the infrastructure itself is secured and trustworthy. In that case, when no adversaries lurk within the nodes to gather information, a way around the paradox of the necessity of transaction data and its use by third parties is at hand.

There are two approaches to ensure that transaction data is not used by third parties: either by strong policies that protect this sensitive data legally, or by an autonomous network, run by actors who do not cooperate with commercial or state agencies.

In the first case, which aims at larger networks, indeed a strong policy to protect transactional data can restrict unregulated transaction data surveillance, or *relational dataveillance*, to some extent. To regulate this issue in general helps to set up an ordered procedure how third parties get access. But legalizing eavesdropping on meta data, from which social networks can be reconstructed (who, whom, when and where) is not convincing within the perspective of a trustworthy network; not really much is gained. Regulation can even cause contrary effects on the practice, as clearly laid down procedures minimize the risk of the surveilling parties to run into legal problems at some stage. And the surveillance case about the Berlin

---

17 The only way to secure it indirectly in an unsecured network is to use the anonymising software Tor. Anonymisers by nature interrupt transactional data processing, or at least make it very hard. See http://tor.eff.org.

activists (chapter 8) has indeed provided convincing insights: even though the authorities had operated with permits by judges for some years, the whole operation was declared illegal years after the fact.

From a perspective that understands institutions and agencies as an ensemble of heterogeneous semiotic-material conglomerations, a lawful interception policy is just one more government program, whose application and practice will show different effects than expected. To promote regulations is equal to asking the state to solve problems. While this is a justified claim for policy scholars, it is not within an analytics of government, where state centrism is not considered providing many insights into *how* government functions. The focus on the practices does not allow for a belief of the black box state: it is one thing what is written down as law and policy, and another what agencies and actors actually do.

Self-Governing Communication Infrastructures

> But I think the way to escape the question with data retention for example could be the idea to completely flatten the idea of what is a service, and what is a user. Because the whole data retention idea depends on that. That there is an ISP, an authority to be asked. But if it's not clear what is the service and what is the client side it will be very hard to track this. And if you run your email server on your mobile phone, it will be even harder practically to do anything about it.
>
> Startx / London

The second case, running an infrastructure of one's own, offers an alternative then, once the operators both take significant steps to secure it against attacks and operate it with an ethics of non-cooperation with state agencies. This sounds rather unusual, but in fact within activists and hacker communities alike, a common sense exists: that the state should not interfere with communication technologies. This has been expressed by most of the interviewees as well.

And this claim fits interestingly within governmental modality of rule, where the state shall not intervene into the flow of things and signs. The difference then of course is here, that none of the activists was referring to commercial players who would organize the field of communication, as in the enterprise society. And this difference is the tipping point where governmental rule has an open flank towards self-determination and autonomous practice: the rejection of state interference is a theme which also circulates in anarcho-syndicalist groups and peer-to-peer production milieus, amongst others. This common sense can be traced back to the demands of ownership over means of production and thus links to an old heritage of work struggles. That neo-liberal rule shares this apprehension to some extent, although under the rigidly different imperative of market forces, provides a point of departure within governmental rule that points towards self-determination and control over communication means.[18]

Materialist Inquiries into Communication Infrastructure

What follows are detailed quotes from one interviewee, Startx, who has spent many years on developing and maintaining non-commercial community wireless networks in London. His account is very enlightening in the matter of autonomous or self-governing infrastructures.

> For me the idea to have control over your network is the first important thing. Because if you have this control you are able to prevent people from surveilling it. What I found fascinating in the last years was the speed with which surveillance moved forward on the internet or any kind of computer technology. If we see how this happened in former levels, like surveillance of normal mail, or landline telephones, we have seen a huge leap forward in surveillance in any kind of computer technology in the last three years. If you see the data retention laws, if you see the questions about encryption: we have to remember that in Britain the law actually could force you to reveal your encryption key to authorities. And what I found fascinating is how naturally people accept these things. If you plug

18 This is were the theory of the multitude connects to governmentality. Post-operaist theory claims production as the central axiom of a theory of politics. "We need to discover the means and forces of the production of social reality along with the subjectivities that animate it." (Negri and Hardt 2000, 22). By widely acknowledging Foucault's importance for a materialist analysis, Negri and Hardt state "what Foucault fails to grasp finally are the real dynamics of production in a biopolitical society" (Negri and Hardt 2000, 28). And of course, their reading of these "real dynamics" is contestable, too, as it resonates within a larger frame of class struggles and contradictions – abstractions that are not present in Foucauldian thought. At the core of this critic, it seems, are remnants of class relations and state centrism induced by Marxism which prevailed in radical Italian thought, although in a very special form, as the central passage these thinkers refer to is Marx' note on the "General Intellect". Form here, many thoughts developed which understand commons based knowledge as the driving force of emancipation. Nonetheless, the focus on the productivity and historicity of subjectivities makes both theoretical lines allies in the analysis of contemporary modes of production and social relations.

some computers together and send some data over it, that is like we talking with each other. There is nobody else involved. What would the people say if there was a law saying: whenever I talk to you I need to call the police and ask to come for a coffee to sit in between us and listen. But people naturally accept it.

Startx / London

This leads him to see the general question:

For me the question is generally: is there any chance in the future that we get back to the idea that communicating via computer technology is as normal as talking with each other. Or if we just give this up and basically accept that everything we do via means of technology is controlled by default. In that case I personally would just drop it and that's the reason why I want to replace the internet. Because personally I don't think there is a chance to get back the internet as a forum for free information exchange.

Startx / London

And he frames it within the trajectory of the current internet development:

And now we go back to this basically cloud computing idea where everything is on the internet, out of control, somewhere far away, nothing is localized anymore. The tendency will be: everything you boot up is a kind of network boot, via DSL. And if you haven't paid your rate, you won't boot anymore. This is a bit exaggerated, because basic access will possibly be free of charge. But if you don't play their rules, you don't boot. And one might not even be able to use the internet anymore. Because you get your internet service from Google. So this whole development goes back to something that would be very easy to surveil.

Startx / London

For the future he sees two parallel developments:

I think for the future there are two ways to handle global networking: one would be what it looks like now, a very centralized approach, of people offering services, and people using the services, and the users are dependent on the people who offer it. And this dependency is commercially interesting. And on the other hand is the question of the potential of people using the computing devices they have and carry around, and if they have access to the devices, and can control it. This would open huge possibilities But there is also the social question if people actually want that. If people are actually interested in the aspects.

Startx / London

And he explains that such a network of mobile device would offer different services than the internet:

> This has nothing to do with the internet as it is anymore. It would be a second network. It is just using by accident the same hardware and the same software, and that also means that you have to change the way you use it and understand it. The interesting thing is that people actually started to do that: with file sharing, people actually do things like that. This is still working on the current infrastructure. And people are actually using that, but they are not aware of it.
>
> Startx / London

The idea of to replace the internet infrastructure came out of a learning process. Before this, the internet was the reference to rebuild.

> Also, I think we made a mistake, we tried to replace the internet partly by offering the people the same services. Like saying: you could do exactly the same internet, by just doing it differently. But I don't think it works in that way. The network structure I described will have to use completely different ways of offering services. And of communicating. To reproduce the same client server-network is not how it will work.
>
> Startx / London

This is what the idea is about: an ad-hoc decentralized mobile network supports and reflects the problems learned about the ongoing centralization of the internet. It can work without a central figure, like a system administrator, who has access to all data, because such a central node is non-existent in such a topology:

> The advantage of the mesh network is that you don't need any authority in the middle to create the network, you don't need anybody to maintain an access point, or decide how the network looks like. The disadvantage is that with current software technology, the routing protocol is relatively unstable, and complicated if it is a big network.
>
> Startx / London

The instability is mainly caused by the routing protocol, which has scaling issues.[19] Thus, a mesh network is currently limited in size.

19 I discuss the differences of protocols for mesh networks with a focus on the B.A.T.M.A.N. protocol in regard to autonomy of networking in Leistert (2010).

> We experimented with different network topologies to create local wireless networks. The two network possibilities are either you have a star shaped network, with an access point where people are connected to, one hub in the middle. The alternative is to have a mesh topology, everybody is connected with everybody in the network, as far as you are in reach. You can actually route all data as far as the mesh reaches.
>
> Startx / London

Startx sums up what such a network means to him:

> So one of my motivations of having this idea of autonomous computing is: can we use long distance networking plus having everything under our control and that as a way to escape surveillance. I wouldn't necessarily limit this on surveillance, there are lots of other things like the issue of intellectual property or simply access, the question of who has control to the network, who controls who is in and who is out. It is not only about control, but if you are actually able to communicate with others in the future. For me, this is the big motivation to do it.
>
> Startx / London

James Stevens, who was one of the first activists who started community wireless networking in London, gives a detailed account about Consume, the initiative that represents these efforts (and of which Startx as well is part):

> The tactic we used with *Consume* was: short circuit the Telecom local loop which governs how people connect to each other, even to their neighbors. You can short circuit that need by providing a connection. Next: make your switch. That's a kind of double meaning kind of statement. Change your behavior, but also practically building a switch. To enable that change of behavior. And three: consume the net, which alerts to a ferocity, which did follow a huge ballooning of file sharing interest between 2002 and 2005. So that was fun. We were self providing. We are continuing to question the dominance of the corporate media with alternatives, not always practical, but they have legs. Up to the point were we are now and there we have a kind of modest network, I suppose, by some standards, 50 nodes facilitating about 250 people being connected all the time. I'd like to grow that. All experimental attempts to bridge out from corporate hegemony. Because it is not a commercial venture in any way, we are not limited by the expectations of our users in the same way. It is also about empowerment. That is part of it. It is been a worthwhile journey to this point, I think. There has

> always been a request for autonomy, and then once you understand the mechanics of networks, then it is a philosophy that you can promote other people to adopt. And how far up the system can you propagate this idea. *Consume* is about that. It was an advocated free network, we were hopeful that we were actually peering as a collective with a conventional service provider. As a network. And negotiating our much needed internet connection. In exchange for access. Once peering had been explained to us it seems that would be a worthwhile ambition. There was a notion to try and use this wireless technology. Broaden the access to the distribution of the internet we had. Now it is everywhere. There is a kind of ubiquity to wireless LAN. There was an anxiety about openness. And freedom, to some extent. You can't always see the benefit in the short term of the position you are holding. Hearing these peoples' responses later on is an encouragement to continue. It is our leisure exploring near future technologies. That might have an effect on everybody else
>
> Stevens / London [20]

In an article about the history of the project, together with second founder, Julian Priest, he writes about the spread of the model: "The model was quickly replicated elsewhere, especially in the States, and there is now a burgeoning wireless freenetwork community around the world" (Priest and Stevens 2004, 125).

In Seattle I had the chance to interview another senior activist of community wireless networks: Matt Westervelt. He recounts the moment of the first steps when wireless networking routers were hacked for the first timeso that later on open source firmware could be stored on the devices, which was a preliminary step for many open source wireless networking protocol designs, now in use by wireless communities around the globe:

> Here we had like the most hardware hacking information than probably in a lot of other places. The link system WRT stuff. We had read that it was a Linux box and they were saying it. And somebody had already strings on it running for some time. Well, if it is Linux and it is from China, they probably did a bad job at it. So we started taking a look. And we figured out that we could ping it. So we asked, how could we get on this machine? We found a testing tool ttcp [a utility program for measuring network throughput, O.L.], and we used that to actually execute code on that machine, so we had a kind of a complicated prospect at that time: you would

20 "Peering" refers to the process that internet providers amongst themselves do. In general, these are negotiations about traffic flows. The protocol for this is Border Gateway Protocol (BGP).

> type it in and you would basically attach it to the web page and then run this Perl script and it would create a reverse shell. It would actually upload code through this testing tool, put a reverse shell onto the box and then log you into it through telnet. And from then there was like: Ah, now we are on the box! We can do Linux stuff with the tools that are there. At that point we couldn't actually write firmware, so we put all the stuff on the Wiki and other people came along and said: ah okay, now that you have got the shell on the box, we know how to do this, and then getting firmware was actually some other project. We were just happy to get the shell on it. From then it was like fun stuff and then creating the distribution for it. The name of the box was Linksys WRT 54g. OpenWRT and all those sort of open projects I can actually trace back to a hacknight at a cafe down the street. That was in 2001, maybe 2002. That was one of the fun things we did here at hacklab.
>
> Westervelt / Seattle

The second step in the history of open source wireless community networking was to produce cheap antennas. Here, again Westervelt remembers:

> People could get that: you take a Pringles can and you stick a wire in it and now you get Wi-Fi [Wireless fidelity, O.L.] from far away. And everyone has done it. And we thought, OK, let's buy a bunch of Pringles. At a hacknight we had maybe a hundred cans of Pringles and we had a lot of stuff and got really sick from the chips. And we made antennas and they are all kind of lousy, but it was fun.
>
> Westervelt / Seattle

As with this antenna, the costs to set up a community wireless network dropped to amounts that made it feasible for low-income people as well. This model became, although not the best performing, a reference object and gadget within these communities.

These most radical proposals and concrete attempts to build an infrastructure from bottom-up for e.g. local message delivery have been a great success not only in London and Seattle[21], but in Berlin[22], Barcelona (were eventually the network turned into an official municipal giant mesh network), and other cities.[23]

---

21 http://seattlewireless.net (accessed 4 March 2012).

22 http://freifunk.net (accessed 4 March 2012).

23 See http://global.freifunk.net for a (naturally incomplete) list (accessed 4 March 2012).

To conclude: surveillance is an issue of the infrastructure. This appears solvable from the perspective of autonomous communication infrastructures, at least on a community or local network scale, as wireless devices can be merged into a network themselves, where clients also act as routers.[24] As much as network traffic is local in fact, for example within one city, the reach of these networks is big enough for many uses.

But in profound ways, these ideas are advanced concepts of network topology and should be regarded as proofs of concept. They echo the history of the internet in a certain sense: while the initial concept as a decentralized topology, and thus a distributed concept of control has changed to a business of very few corporations[25], radical developers, such as James Stevens and Startx, are envisioning a return to decentralization or distributed systems even as a response to pressing issues of surveillance or high prices for simple services, such as SMS. By setting up wireless networks that cover large parts of big cities, based on protocols that encourage distributed topologies, a bottom-up desire responds to rather irrational costs for local mobile phone communication.

Autonomous Communication Means

One approach to reduce the spreading functional creep and misuse of transactional data in unregulated environments are strong privacy protecting policies. But this cannot solve the problem at hand: the infrastructure remains in control by entities which, by way of controlling it, can exercise power on its users; additionally, a regulated surveillance scheme integrates surveillance into orderly rule and makes it a common practice. On the other hand, the radical approaches of a self-controlled, non-commercial, autonomous infrastructure offer many prospects to the societal dimension: no central instance decides about who can take part and who cannot. No discrimination on the content of the data can be set in action, like traffic shaping. Finally, meta-data cannot be collected and analyzed by a centralized actor.

But there are clear limits, both in the technical and in the social dimension.

24 Dunbar-Hester (2009) discusses how activists negotiate FM and Wireless spectra use and the engagement with DIY infrastructure technologies.

25 The discussion on net neutrality echoes the concentration of power in the internet domain very well. To prioritize one kind of data over another one is the harshest intervention in the history of the internet so far and sets forth a trend of commercial take-over.

Technically one should consider these approaches at the most as community projects. They are not yet capable of offering a quality of service that is comparable to the 'internet'. This seems a real problem, as it remains to be seen if users are willing to decrease their network expectations for a self-governed infrastructure.

Socially these approaches are challenging, because they rely on an activation of users in a governmental sense: to conduct them to conduct themselves into networked autonomy. The benefit of a lesser degree (I hesitate to write 'no') of surveillance is an abstract good for most network users. The benefit, that traffic cannot be discriminated on the base of its contents, like in commercial DSL, where peer-to-peer protocols get less bandwidth than a stream from a commercial content provider, can be more convincing. Sharing of data without interference by state or commercial actors might be a mass popular feature once the idea to build such infrastructues on a community level has been sufficently communicated.

Recalling the first part of this chapter, and combining it with the second, the sum is ambivalent: although means are available to escape the surveillance scheme in networks in general, one is automatically forced to enter into a mode of communication that might never be adopted by the masses. By fragmenting the physical networks, also the entire sphere of communication gets more fragmented.

10 Conclusion: Mobile Media Pushes the Liberal Paradox

> There is a group called Maxakalí. They are indigenous people that have been in contact for 300 years, but remained monolingual, almost no women speak Portuguese, only men. They are a poor community, very little resources, on the northern boarder of Minas Gerais. The land they have is poor and little, the forest has been destroyed. Their basic material is clay. And plants. They have a ritual there where they build clay mobile phones. And they use it to communicate with the spirits. Although they don't have mobile phones, through this ritual they acknowledge the potential of agency that is in that kind of technology. They are trying to relate to it through this ritual. This is real, not "as if".
>
> Caminaghi / Campinas (Br)

The intention of this work was to show, that a discussion of mobile media as a form of government, as part of governmental rule, contributes to an understanding of how power relations are reshaped and invoked within a conglomerate, that can be addressed as a political rationality. A rationality that invokes practices which are guided by programs of government.

The discussion was informed by many diverse statements from an geographically even more diverse set of experts, whose statements translate a discourse about mobile media, and, specifically in its empowering mode, for protests, and, in its surveillance mode, for a modification of sovereignty to neo-sovereignty.

The easement of communication amongst the population fosters productivity, and, as such, the roll-out and massification of mobile media is a technique of a governmental scheme. It has become incredibly tedious to not be connected.

> Also, once you use the mobile, you can hardly get out of it
>
> Kumar / Delhi

This signifies that an understanding of mobile media can not neglect its enmeshment into power relations: it conducts to a self-conduct of 24/7 connectivity. The machinic space of mobile media provides an addressability, both socially and technically. Parallel to the setting up of a postal delivery dispositif, this mobile media dispositif connects very large parts of populations globally, and addresses the individual at the proximity of its body. There remain resistances against such an intermingling of body and technologies for less governmentalized subjects:

> I have suffered from surveillance. Communication is a kind of limitation. You want freedom for a day. And if you have a phone, you don't have freedom because people can reach you, you can not loose yourself. Anyone will find you. This is a problem. Everybody wants freedom, but nobody wants to loose access.
>
> Anonymous People from Ankur / Delhi

This liberal rule via mobile media triggers illiberal governing that translates into neo-sovereign rule. Here, the immanent paradox of liberal rule manifests itself: it needs to secure its production of freedom, but at the same time these operations of security not only provide a milieu of circulation, but intervene; thus, the paradox. The production of freedom shall not intervene into the transit realities of rule, but does so constantly. Invocations of data retention, blocking of communication, and a mandatory registration of SIM cards can be seen as securitization techniques, that are both, invoked by liberal rule and in contradiction to its imperative of the flow of signs. Mobile media thus provokes the liberal paradox to the extend, that it becomes regulated, bound to territory and scrutinized. I termed these operations "neo-sovereignty" because these are reconfigurations of interventions in the style of sovereign rule, but within a neo-liberal paradigm.

The machinic addressability, it seems, provokes these technologies of control, as digital surveillance schemes are cheap, and provide the production of a new kind of knowledge: about the population and the individual alike. An emergence of algorithmicly produced digital objects, which do not *represent* the target, but are a *site of production* of its double; a fragmented relational set of flecks. Government, which calculates its operational knowledge on these technologies, and decides about measures and programs on such doubles, manages abstract data, which lacks ontological resemblance about the governed. Or put better: the new ontological foundation through such means remains unproblematized.

Securitization techniques are taking advantage of mobile media. Data retention is a technology of securitization with enormous powers, and today we can only understand the tip of this iceberg. Data retention is a gigantic illiberal materialization of a doubt: a doubt that is large enough to hold the whole mobile population. Mobile media thus inscribes itself not only in governmental programs and is one in itself, but changes governing in general. When it is invoked by securitization, one may speak easily of mobile surveillance media.

On the other hand, mobile protest media supports in many ways the aims and ambitions of protesters. New forms of distributed actions have emerged, that can tactically overrule surveillance. But there are frictions between the collectivity of performing dissent and the powers of an individualizing technology.

There is also dialectic at work here. Surveillant capacities, programs and technologies adjust to the increased agency of protest groups, or likewise, the increased mobility of populations. The trajectory of neo-liberal programs cannot but push for technologies that ease and increase the flow of signs, goods, and people. In its paradoxical fold, neoliberalism then invents illiberalizing programs and technologies.

> In the context of the social repression here in Oaxaca, after the repression of November 25th 2006, which was the most violent attack of the federal police downtown, human rights activists were trained in the use of mobile phones to try to contain the surveillance because the surveillance was declared as a permant reality. So, we can not fight it, we can not stop it, but we can try to contain it and we have to life with it. The use of the phone, not saying a lot, not saying delicate parts of your life
>
> Blax / Oaxaca (Mx)

An interesting example for such an excess of illiberality, and a signifier of what is on the horizon of mobile surveillance media, is a product[1] recently acquired by the Metropolitan police in London. The product's capacities are explained in a sales leaflet."Geotime" promises a new operationability of real-time mobile media surveillance:

> The Metropolitan police has bought Geotime, a security program used by the US military, which traces and displays individual's movements

1 http://www.geotime.com/Product/GeoTime-(1).aspx (accessed 6 March 2012).

> and communications with other people on a three-dimensional graphic. It can be used to collate information gathered from social networking sites, satellite navigation equipment, mobile phones, financial transactions, and IP network logs.[2]

Whatever difficulties this product in reality may encounter, and if it works like it is portrayed, is not of concern (it would require empirical research). The dispositif it represents in the fantasy of a sale's brochure allows for a speculation: mobile media's political rationality *itself* is under negotiation, under rearrangement and constantly under pressure. It is not a solid floor. Media technologies themselves co-develop with political rationalities and vice versa. In this interplay, different strategic ensembles produces historical frames, but each time frame is under tension, provided it unfolds within the liberal paradox.

A Development Model of Media

Finally, there is a speculation possible, which targets a development model of media, that I want to touch briefly: one can look at this articulation of material/technology out of a discontent (like Startx stated: about the development the internet has taken) from an angle that Hartmut Winkler proposes:

> If the development of media rests on desires [needs], and these are themselves co-determined by the state of history/media history, then discourse, before it brings into existence the next step of media technology, first of all seems to produce something else: some negativity, a discontent, some kind of deficit, a system tension, this very 'desire' [need], which is a prerequisite for the very next step of media development.[3] (2004, 142; translation O.L.)

2 "Police buy software to map suspects' digital movements", Ryan Gallagher and Rajeev Syal, guardian.co.uk, 11 May 2011, http://www.guardian.co.uk/uk/2011/may/11/police-software-maps-digital-movements (accessed 20 March 2012).

3 "Wenn die Medienentwicklung sich auf Bedürfnisse stützt und diese vom jeweiligen Stand der Geschichte/Mediengeschichte mitbestimmt werden, dann scheint der Diskurs, bevor er die nächste Stufe der Medientechnik hervorbringt, zunächst etwas anderes zu produzieren: Ein Negatives, eine Unzufriedenheit, eine Art Defizit, eine Systemspannung, jenes 'Bedürfnis' eben, das Voraussetzung für den jeweils nächsten Schritt der Medienentwicklung ist."

Recalling the heavy instrumentalization of mobile media for neo-sovereignty, the system tension is apparent: what initially was announced as liberation, more and more turns into illiberation, an actual threat for the user.

> This negativity appears to emerge unintentional, at the back of discourse; and if at the same time it is the *motor* of development, it is less important what a discourse produces positively, than what it does *not* produce; what is left open and impossible to state [say] under current conditions of media.[4] (Winkler 2004, 142; emphasis in original; translation O.L.)

What has become impossible to communicate in mobile media environments are statements that rely on privacy, and what has become impossible to hide, are one's social relations and location. Once can say that a historical phase is over: that little historical window of liberalized (or anti-sovereign) mobile media which emergeed between the massive roll out of mobile media and its excess in illiberalization through measures of securitization. The window of opportunity for mobile media to operate freely from the state is closed. In this sense, the joint efforts by Startx and others in the field of alternative networking infrastructure echo that "the development of technology in this sense plays shots."[5] (Winkler 2004, 142; translation O.L.)

The autonomous network solutions, whicht Startx referred to, are designed explicitly to connect mobile media not via GSM, but wireless LAN. Telephony via wireless LAN is a popular service, as it is much cheaper than telephony via GSM, and it can rely on infrastructure which is not controlled by large commercial players. One of the Korean anonymous activists pointed to internet telephony:

> When you talk about mobile communication, I think it is very broad. So you have cell phones, laptops, mobile internet, but personally I don't like it a lot. I don't have a cellular phone myself, I have an internet phone, a wireless phone, which works were there is wireless LAN
>
> Two Korean Activists

4 "Dieses Negative scheint ungewollt, d.h. im Rücken des Diskurses zu entstehen; wenn es gleichzeitig der *Motor* der Entwicklung ist, wäre weniger wichtig, was der Diskurs produziert als was er *nicht* produziert, was offen und unter den gegebenen Medienbedingungen unsagbar bleibt."

5 "Die Technikentwicklung spielt insofern immer über Bande."

He showed me his phone: it looked like a cell phone, but would work only via internet connectivity. For Seoul a clever solution: the city is covered completely by wireless LAN.

Appendices

A Empirical Research about Protests: Ethical Considerations

Research about a presumingly progressive use of mobile media technologies and its ambivalences in regard to surveillance and repression can produce a fair body of knowledge without empirical inquiries. But there is a natural limit then: it remains speculative in big parts. Whatever sources of information are used, they are never able to give an account of concrete, situated practices. A lot can be done without interviewing people, of course.

When I started to conceptualize my research, it was of importance for me not only to do theory, to reflect on other research, to develop my own theoretical puzzle pieces, but also to do empirical research, better said: qualitative research that mirrors discourse. I was never satisfied with the speculative nature of a lot of work on media practice in general and about the specific use of telecommunication and mobile media in the context of social movements and individuals fighting for social justice.

When it comes to social movements though, the situation is complicated. Empirical research here encounters by nature difficulties and often remains ambivalent. Questions of trust, of goals, of ethical conduct are arising. Why should people within social movements talk to researchers? What do *they* gain? Spending precious time with some researcher whose intentions remain mostly vague for the interviewee is not the most appealing activity.

One of the reasons for this reluctant attitude is fear of repression. A lot of my interviewees or their acquaintances have faced or are facing repression, or simply know about the possibility thereof.

To make plausible what the ethical problems for a research in this field are, it is instructive to give an example: the social movements on the Philippines are strong and impressive. And they are working in a very hostile environment. All of a sudden people are going missing, are found dead, are being harassed, or are facing threats. Especially those that are publicly addressing human rights issues and corruption are living a dangerous life. I talked about this with Jun Cruz Reyes, a well known Philippine writer.

> My writing is about the hypocrisy in the Philippines. It is about power relations. The Philippines is a third world country. The establishment is very much insecure about their position in power. The president has given order to eradicate insurgency activities by 2010, I assume I am part of the deadline also. Anyway I am still saying nasty things.
>
> Reyes / Manila

We met on the campus of the University of the Philippines. It was only very recently that he had been harassed by most certainly the military. He decided not to stay at home, but to live on the university campus, where he felt more secure. Although not participating in a political group in a strict sense, Reyes has a network of people to support him – including journalists. The mobile phone then played a key role to mobilize this network for support. Very fast and effectively Reyes could organize press conferences about the harassment and make it public.

> Just two weeks ago, I gave a press conference, because I had been harassed by the military for what reason I do not know. I guess it is about my writings. And I was surprised: media people, newspaper, internet, radio came there. I also heard that my readers are complaining as well. I used my personal phone to call three or four of my media friends, and they spread it further. It was even a holiday, over the weekend.
>
> Reyes / Manila

As Reyes is publicly scandalizing the practice of counter-insurgency politics and police/military action on the Philippines as a method to silence critical voices, for him it is not counterproductive to talk to a German researcher about it.

Filtering Statements

I am confident that my interview partners all had well considered what to say or what better not to say in the interviews. During transcription though, it appeared to me that numerous statements should not be published. Some statements did not contribute to the subject of this research, others were sensitive enough to possibly cause troubles. Of course this is a thin line: it remains a speculation if it would cause trouble. My guideline to decide whether a statement is unproblematic or not never became completely transparent to myself. Guided by long years of media activism practice that was focused amongst other things on security of telecommunication for social movements, I decided not to use roughly 20% of all statements I collected. I even omitted their transcription to forget about them myself. Otherwise I would re-encounter these statements every time I reviewed my material. The decision not to use them would otherwise be questioned again and again: an exhausting process. From the standpoint of a researcher this loss of material is bitter. Some of my interviewees also asked me to review all statements once they were written down prior to allow me to use them. This made it a tedious process to build a base of usable material.[1]

The central questions that needs further reflection for this research in regard to benefits and dangers for the interviewees therefore are:

- What is this research for?
- For whose benefit is it done?
- What are the (mostly uncontrollable) possible side effects?

These questions are addressed in the following paragraphs.

What is this Research for?

In many if not most cases academic research rests in academia – for many and various reasons. If a general statement about this can be made at all (and I only touch this question briefly here), it targets the specialization of academic knowledge production. As most things in modern life, it is a specialized task by highly specialized individuals within a specialized community. It is for this reason hard for a researcher to produce repercussions in the non-academic world. For some researchers this setting does not cause problems; some might not even regard this as a deficit of their work. And for sure, not every research directly addresses issues

---

1 One interviewee did not respond to my transcript. I did not use the interview.

in this sense. But this research wants to produce knowledge that is applicable both inside and outside academia and both in different ways.

Regarding academia, one can try to weave a work that takes part in contemporary theoretical and empirical discourses on the research's subject and related fields, visit conferences on related subjects and get into a dialogue with researchers that work in the same areas..

Regarding the non-academic world, the case is different and surprisingly much more complicated in a lot of ways. The borders one builds by writing an academic work often make it hard for non-specialists to follow terminological pathways. While my general idea from the beginning was to contribute with this reflection to the activities of the interviewees, the ethical urge had been drastically raised and been concretized after the 50 interviews I did. The impressive knowledge and experiences that have been expressed and that I was able to collect should not remain isolated in few academic libraries – it should find its way back. I hope that in this or that way it does find its way back and offers a reflection and critque of the work of activists from a specific perspective that comes from outside. The knowledge reproduced here is original and as empirical research it is combined with theoretical and discursive reflections, leading to a synthesis and hopefully a constructive critique. The outcome – this is my hope – can be informative for any groups that are working towards social justice with means of mobile media.

Language

Writing in English was not an easy, but most coherent decision: it still remains the only viable way to have a possible reach beyond most language borders. My research travel made me realize how problematic the general assumption about English as a world language is. It is by far a higher class language in a lot of countries, especially in post-colonial countries with a Commonwealth history. Speaking English is speaking to those that have *learned* it because of access to education, if not higher education. And then one speaks the language of the former colonizers.

English, like any other language, transports concepts, mindsets, values, and belief systems. These come from Western, Christian culture. Furthermore, English, because of its imperial history and contemporary function in a global capitalist economy, is not the most preferred language in social movements outside native English speaking regions. I encountered harsh reservations against it, mostly in

Latin America. Therefore the ability to speak English was not automatically leading to actually speaking it. And also, what one is able to say in English differs widely from what one is able to says in one's everyday language like Kannada,[2] for a variety of reasons. This starts with the term 'surveillance' itself which lacks proper translation in many languages. Missing vocabularies are only one problem. Speaking the language of former rulers (in the case of India, Pakistan and Bangladesh, but also the Philippines) is also a different emotional experience than speaking the common regional language.

For all interviews I did with people that did not want to conduct it in English, translators carefully tried to transpose all statements into English. By choosing translators which in most cases had a comparable background, i.e. having been social and/or political activists themselves or having been close to those who are, I tried to make sure that the contexts of the activities that the interviewees were reporting about did not get lost. Otherwise I would not have been able to trust the translations to the extend necessary for this work. I still feel indebted to everyone who translated.[3]

The method of translation was bound by the method of interviewing: a guided interview follows the interviewee to some extent, without forgetting its goals. Therefore it was not a simple matter of translating the question and then the answer, but also sustaining a lively conduct. The artificial breaks caused by translations are dangerous moments during interviews as one side then is excluded and easily looses track. A help is to keep eye contact with the interviewee even without understanding a word she says.

In a lot of interview situations, I encountered another dimension of language issues: English in India is something very different to English on the Philippines or in the US. This leads to English words which emerge only in one region or country. Accents and intermingling of English with regional languages sometimes gave me a hard time to understand. In many cases, though, I could resolve issues during rehearsal and transcription. Sometimes I had to contact people during transcription

2 Kannada is one of the dominant languages in the Indian state of Karnataka of which Bangalore is the state capital. It is spoken by around 38 million people.

3 Thank you: Olivia (São Paulo), Minerva (Mexico City), Akiko, Kaori, Hajime (Tokyo), and Dinesh (Bangalore). Due to the nature of the Lawyers' Movement, there was no urgent need for Urdu, Sindhi or any other translation in Pakistan. Everyone I interviewed in Pakistan fluently spoke English.

by email to make sure I got it right. And in very few cases, I am still unsure. These passages are left out of the work.

While some interviewees spoke very eloquently, others did not. This is still present in the statements. For a better reading experience, I slightly intervened into the words during transcription. These minor interventions though did not change more than interjections or repetitions. As with any audio transcription, punctuation was a decision during writing.

Related to the issue of language is the issue of dissemination. The traditional way of publication as a book has its known limits in practical reach. It is unlikely that my research would be present anywhere else but in very few academic or institutional libraries. The high costs of book production in relation to the financial means of most of the actors whose activities are addressed here are really standing apart.

But there is a way around: the production of an electronic document in parallel that can be disseminated online. Although internet connectivity is not ubiquitous and having such a document on the screen definitely not a pleasure in comparison to having it on paper, it allows for the interested to have a look and decide if a print out (in whole or in parts) is worth the effort.[4]

Possible Side Effects

The possible side effects of this research are foremost only *possible* side effects. They belong to the same realm as one theme of this research: surveillance. But nonetheless they are worth mentioning. This naturally is a speculation.

The relevance of safety measures is related to the subject's societal environment and position. Many interviewees are underprivileged and vulnerable. Therefore the danger of repression is nothing that can be generalized very well. It depends on the local situation.

it remains the possibility that statements given here are offering hints or contributing directly to open files of authorities, military or other potent antagonists of the interviewees or their initiatives and movements. Because communication methods, tactics and strategies, experiences and stories of successful or non successful struggle aided by telecommunication means are referenced throughout this

4 An earlier version of this text can be downloaded from http://digital.ub.uni-paderborn.de/hs/content/titleinfo/555530.

work, it is always possible that some important hint is given that might be used in a counter strategic way or even direct repression. On the other hand, as I describe the surveillance practice of Law Enforcement Agencies, this is also a contribution to transparency in a very opaque field.

Gatekeeper

In many cases, my contacts to the interviewed people were made possible by locals, which generously offered to help me. These actors are traditionally called "gatekeepers". They are local people trusted by local people. They accompanied me to the interview places, sometimes translated, and sometimes even offered their hospitality by letting me stay at their places. These people made sure that I got to know local contexts and understood better the accounts given. They often provided me with further reading material or hints where specific places would be that have been places of struggles. This was an important contribution to safeguard myself and the interviewees: myself, as otherwise I would have gone lost easily in places like Mexico City, and my interviewees, as the contacts had a sensitive mind on what might not be suitable for publication. They did not intervene into the interviews as such, but afterwards often explained what they had noticed during the interviews. As they were not the persons giving accounts, they had enough distance to be sensitive for such things, whereas the interviewees in their sometimes very detailed and long answers tended to forget about the interview situation and thus security matters.

Anonymity, Pseudonymity

I offered every interviewee anonymity or pseudonymity. Many took advantage of this. While it is of course more coherent to be able to reference every interviewee by full name, the offer for anonymity had to be made, otherwise some interviews would not have taken place at all. Mostly in Latin America and the Philippines people took advantage of this. I do not indicate where a pseudonym is used or a real name.

B Overview of Interviews

When and where the interviews took place and informations about my travel itinerary are listed here.

But first a note about the reasons why I interviewed these specific actors and not others:

I had developed criteria that guided the process to find interviewees: first, all interviewees should in any way understand their work and that of their groups or organizations as aiming towards social justice. Social justice here entails such things like fights for land to grow food, access to knowledge and information, free public transportation, free housing, free speech, emancipation of minorities. It is clear that each of these topics would need a proper framing. I neglect this here. My intention is basic: all interviewees target a more just distribution of signs, goods, and services, one which is not (solely) based on the mechanisms of the "free market".

Second, political parties and unions or other hierarchical organizations are not part of the discussion (MST is an exception, though). This has to do with a hypothesis that to understand mobile protest media, hierarchy obfuscates its power vectors more than non-hierarchical forms of organizations.

Third, my aim to show how mobile media reconfigures many societal fields globally, in very different cultures and regions, made it necessary to interview activists from different areas of the world. Of course, not all regions are present. For example, no African country.[1]

Fourth, the network of people one is part of helped to find interview partners. It took a year to prepare the travel and organize bits and pieces.

---

1 The uprisings in Tunisia, Egypt and other countries started after I had finished my empirical phase of research. None of them is covered here.

Fifth, I was not particularly interested to interview technologists amongst the protesters. Although I decisively searched to interview some free software developers, that count themselves as part of a protest movement, as they would provide an important expert angle on technologies – to gather insights into a discourse about protest and surveillance, it would have been problematic to address technology experts prominently. Thus, the largest part of the interviewees is not active in technology development and can be considered as users of mobile media.

Sixth, while the dynamics of technology development are hilarious, and, for example, 'smart phones' are deployed massively in many countries, I did not focus on these media technologies. Telephony and SMS are the services discussed here most prominently. While it certainly offers interesting debates, the access to internet enabled smartphones has not yet reached the poor ones. Thus, a focus on such technologies excludes large parts of the populations from the start. This is different with simple phone technologies.

The 50 interviews conducted for this research were all audio recorded on site. I used a MP3 recording device, afterwards I transfered the files onto my laptop.

Table B.1 shows the itinerary of the main research travel I did. Before the main travel, I visited London for interviews, and afterwards I visited Madrid for interviews.

Table B.2 to B.5 give a brief overview of the interviews taken and used in this research. Groups and memberships stated are not exhaustive. Not all reports of mobile media practice automatically refer to the affiliations mentioned in the tables. It is very common amongst some of the interviewed people to engage temporarily in other groups or to engage in temporal campaigns. Nonetheless, this table represents a general overview. I have added a column *Translated* which shows where I needed support of translators, otherwise it was conducted in English. *Place of Activities* is occasionally only vaguely stated as some interviewees have a wide outreach with their activities as they are engaged in technical issues of activist communication. Therefore the mentioned places of activities are not always very precise, but nonetheless give an impression. *Active with* refers to present and past. Information in curly brackets signifies affiliations which played only a small or no role in the interview, although they might play a decent role in the interviewee's overall activities. Some interviews are group interviews, from two to four people.

No.	*Place*	*Arrival*	*Interview*
01	São Paulo	24.7.	04-09
02	Campinas	3.8.	10
03	New York City	7.8.	11-13
04	Seattle	15.8.	14-15
05	Mexico City	25.8.	16-19
06	Oaxaca	2.9.	21
05	Mexico City	4.9.	20
07	Tokyo	8.9.	22-28
08	Manila	27.9.	29-33
09	Bangalore	8.10.	34-38
10	Delhi	17.10.	39
11	Lahore	20.10.	41-47
12	Delhi	31.10.	40-41

Table B.1: Travel itinerary 2009. The 6 interviews conducted in Europe are not listed.

Where appropriate I separated the cited sentences and referenced them with individuals, but mostly I did not differentiate the speakers in such cases. For acronyms of groups, please refer to the glossary.

No.	*Name*	*Date*	*Place of Interview*	*Place of Activities*	*Active with*	*Trans-lated*
01	James Steven	6.4.	London	London	Consume	n
02	StartX	6.4.	London	London	Consume	n
03	Yossarian	6.4.	London	London	Indymedia London	n
04	Alface	29.7.	São Paulo	São Paulo	CMI, Forum Centro Vivo	y
05	Legume	29.7.	São Paulo	São Paulo	MPL, {Passa Palavra}	n
06	Bispo & Christoph	30.7	São Paulo	São Paulo	Cooper Glicerio	y
07	Three MST Worker	31.7.	São Paulo	São Paulo	MST	y
08	M	1.8.	São Paulo	São Paulo	MPL, {Punk Movement}	y
09	Fernão	2.8.	São Paulo	São Paulo	Sarava, MTST	y
10	Francisco Caminaghi	6.8.	Campinas	Campinas	{Radio Muda, Sarava}	y
11	Jon Goldberg	7.8.	NYC	USA	{Jews against the Occupation}, I Witness Video,	n
12	Daniel Kahn Gillmor	11.8.	NYC	USA	Mayfirst / People Link, Tachanka	n

Table B.2: Interviews 1 to 12 at a glance, year of conduct is 2009.

No.	*Name*	*Date*	*Place of Interview*	*Place of Activities*	*Active with*	*Trans-lated*
13	Nathan Freitas	12.8.	NYC	USA, Tibet, China	Free Tibet and Anti-Censorship initiatives	n
14	Matt Westervelt	24.8.	Seattle	Seattle	Wireless Seattle	n
15	Devin Theriot-Orr	24.8.	Seattle	USA, Pakistan	Lawyers' Movement	n
16	Anonymous	28.8.	Mexico City	Chiapas, Guatemala	Independent grassroots media	n
17	Enrique	29.8.	Mexico City	Mexico City	Laneta	y
18	Olinca	1.9.	Mexico City	Mexico City	Laneta	y
19	Alpha	1.9.	Mexico City	Mexico City	{Radio}	y
20	Francisco Cerezo	4.9.	Mexico City	Mexico City	Comité Cerezo, Revuelta	y
21	Blax	3.9.	Oaxaca	Oaxaca	Indymedia Oaxaca, Video production	n
22	Yasudo Yukihiro	19.9.	Tokyo	Japan	Labornet Japan, Champon, {MediR}, No G8 2008	y
23	Ahmed Mahmoud	22.9.	Tokyo	Bangladesh	The Voice Bangladesh	n
24	Two Korean Activists	25.9.	Tokyo	Seoul	Korean media activist network, Yongsan Candlelight Media Center	y

Table B.3: Interviews 13 to 24 at a glance, year of conduct is 2009.

No.	*Name or Pseudonym*	*Date*	*Place of Interview*	*Place of Activities*	*Active with*	*Trans-lated*
25	Freddie	25.9.	Tokyo	Hong Kong	Hong Kong independent media	n
26	Hosaia Yuki	25.9.	Tokyo	Japan	No G8 2008	y
27	Higuchi Takuro	26.9.	Tokyo	Japan	No G8 2008 Infotour	y
28	Matsumoto Hajime	26.9.	Tokyo	Tokyo	Shiroto no ran, community activism	y
29	Randy Nobleza	1.10.	Manila	Manila	Indymedia Manila	n
30	Jun Cruz Reyes	2.10.	Manila	Manila	Writer against Corruption	n
31	Non-Collective	2.10.	Manila	Manila	Food not Bombs, discussion group	n
32	Ederic Eder	3.10.	Manila	Philippines	TXTPower	n
33	Joan, Julie, Mina, Minerva	3.10.	Manila	Philippines	Sinagbayan, Linangan ng kulturang pilipino	n
34	MānasaSarōvarā	14.10.	Bangalore	India	Diverse Non Violent Direct Actions	n
35	Christy Raj, Kannan	15.10.	Bangalore	Bangalore	LesBIT, Sangama	y
36	Akkai	16.10.	Bangalore	Bangalore	Sangama	y

Table B.4: Interviews 25 to 36 at a glance, year of conduct is 2009.

No.	*Name*	*Date*	*Place of Interview*	*Place of Activities*	*Active with*	*Trans-lated*
37	Al Alegre	16.10.	Bangalore	Philippines	Foundation of media alternatives	n
38	Leo Saldanha	17.10.	Bangalore	India	ESG	n
39	Madhuresh Kumar	19.10.	Delhi	India	Delhi Solidarity Group, NAPM	n
40	Ankur People	3.11.	Delhi	Delhi	Ankur	y
41	Vijayan	3.11.	Delhi	Delhi	Delhi Forum	n
42	Mohammad Azhar Siddique	21.10.	Lahore	Lahore	Lawyers' Movement	n
43	Ahmeed Salemi	25.10.	Lahore	Lahore	SAC	n
44	Abeer Hamid	25.10.	Lahore	Lahore	SAC	n
45	Hamid Zaman	27.10.	Lahore	Lahore	CCP	n
46	Sakham Kahn	27.10.	Lahore	Lahore	CCP	n
47	Asad Jamal	28.10.	Lahore	Lahore	Lawyers' Movement	n
48	Anonymous2	3.12.	Madrid	Spain	Hacktivism	n
49	Iokese	4.12.	Madrid	Spain	Hamlab / Hacklab Madrid	n
50	Ricardo Dominguez	4.12.	Madrid	Borderlands	EDT, B.A.N.G.lab	n

Table B.5: Interviews 37 to 50 at a glance, year of conduct is 2009.

C Index: Groups and Initiatives of the Interviewees

The main trajectory here is not to research social movements, initiatives and groups themselves. Nonetheless, it is still helpful to give an account of the interviewees' activities. Otherwise the accounts about their mobile media use would often lack context.

What follows are descriptions that dominently are provided by the interviewees themselves. I left out the groups and initiatives which do not play a role for the mobile media practice the interviewees spoke about. As often, the accounts given did then refer to other activities which developed parallel to the main engagements and activities, but not within the main group of the interviewees. Where necessary, I added further explanation. The general goal is to keep it brief. It is not primarily intended to be read in one part, but as a reference to look up. As I hesitate to categorize the groups by categories, they are ordered by country and then alphabetically.

Additionally it is important to note that some interviewees are not present in the following list, as they either did not count themselves part of steady group or movement or they did not refer to a group strong enough for me to indicate the relevance to name it.

Brazil

CMI and Forum Centro Vivo (São Paulo)

> We do a lot of communication. That's why I am part of Indymedia, CMI. I am mostly engaged with social movements here in the center, downtown. Which means my involvement is with the squatters, housing movement, the garbage collectors, who do recycling, with informal street vendors, street dwellers. Everything that is poor.

> All that is a outcast. And we have a forum, called *Forum Centro Vivo* and in this forum we try to work together. Because every movement has its own focus. The idea is to have a common agenda. And you have people from universities working together with social movements. And a lot of NGOs, from the church to whatever.
>
> Alface / São Paulo

The Brazilian Indymedia is called *CMI*. Many of my interviewees from Brazil count it as a major source of information about activities and grassroots movements in Brazil.

> Since last year with the state government, we are mostly involved with evictions. Of course there are a few possibilities to occupy something, but very few here in the center.
>
> Alface / São Paulo

Occupations and squatting in São Paulo are also the background of MTST and the two interviewees Fernão and M.

Cooper Glicero (São Paulo)

The *Cooper Glicero* is an effort of self-organization from poor city dwellers and trash collectors to raise their standard of living and to organize themselves.

> The activities involves strengthening the struggle of the collectors, organising the collectors, so that they can claim collectively the acknowledgement of their job. And claim rights, be acknowledged by government and society
>
> Bispo and Christoph / São Paulo

As the workers are spending their work day throughout the city, they are relying on mobile phones to have internal communication in case of emergency situations or aggressive behavior towards colleagues by authorities.

Movimente Passa Livre (São Paulo)

> From the free transportation movement, we think we have to change the idea what a city is. We think the city is a right, and not only a thing to consume, but something to have access to. We fight that others have the right to use the city. To use and to change the city. We use the same concepts as David Harvey has for the city. So we fight for a different city that is more participative for the people.

> The public transportation in Brazil is very low quality and we fight to get a better public transportation, but not only that: we think the city discussion is very important, too.
>
> Legume / São Paulo

David Harvey is a geographer coming from a Marxist tradition, who is an important point of reference for numerous anti-gentrification activists and people who are trying to prevent urban space from being privatized.

> For public transportation we have so many things: we made manifestations in the streets, we do work in secondary schools, we made a discussion about transports in the schools with teenagers. We let the people think about the importance of transport and their own life. We have some activists that do workshops and discussions that try to go deeper in discussion, more deeper thinking about city, and we have these discussions in communities away from the centre, in suburbia, about transport and about the problems of these communities. And we have articulations with different social movements, like MST and MTST,
>
> Legume / São Paulo

M is focusing on the involvement of the youth in the free transportation movement.

> The Free Pass Movement comes from a dissatisfaction, mainly from students in the public high school system. These students were dissatisfied that the student unions were used by parties. Basically it startet by 14 to 18 year old people, influenced by popular revolts against parties, e.g. Trotzkists. And they reclaimed autonomy, self-organisation, this kind of claims. When the movement started in 2005, it was from dissatisfaction from the organised left in Brazil, and also because on the other hand there was this problem that cities were growing larger and especially youth wouldn't have access to cinemas and such things that would happen in the city, football games. As they were obliged to have access, there was nothing to be done in the neighbourhood. In short, the Free Pass Movement defines itself as an autonomous, non-party, and horizontal social movement for public free and good qualtity public transportation. We believe it is a right to the city, it should not be paid for. Therefore we have good connections to the homeless movement and the landless movement. Our main actions usually happen when the fares go up, then we can gather more people. It is a national movement, demonstration happen nationwide. And when the fare has gone up, we have been able to gather some 5000 people.
>
> M / São Paulo

Movimento dos Trabalhadores Rurais Sem Terra (Brazil)

MST is a huge organization with branches all over Brazil. It has a lot of power as it can mobilize hundred thousands of people. It is hierarchically organized, has paid workers, and even an alternative MST University. MST positively refers to the Cuban model. This makes it the most traditional leftist political movement in this research.

> MST has national congresses every five years, since the movement was founded in 1984. Every five years there are goals defined, some are permanent. 1. the fight for land. 2. the fight for land reform. 3. the fight for social transformation. Every five years, new things arrive, then we adjust to these specific historical moments. Fight for social transformation: MST comes in a context of prior struggles. Land is a major issue here, through the concentration of landownership in Brazil. Therefore we believe that the struggle for deconcentration of landownership implies the deconcentration in many other power fields.
>
> MST Workers / São Paulo

MST is organized in specific branches, whereas the fight for land and land reform is its dominant part. The worker here recount the activities in all their diversities, up to international relations.

> The movement has different work divisions. First, access to production, that involves mode of production, use of toxic stuff. 2nd is education. MST has always worked on education. MST says we want to break the case for land and education. Knowledge and school are as concentrated in Brazil as land. This agenda has become more complex in the history of MST. Education know means working in literacy for every age, a sector for political education, small children and a programme for access to university. Another activity is communication, that exists since the beginning of the movements, where we already had a news bulletin. Two main needs for internal communications. MST is the first national land issue movement, a pioneer in landmovement, for its geographical scope. For internal and external communications, we have different media: news bulletins, community radio, the website, pamphlets. More sectors are: in every sector is a discussion of models. So, in the health sector, we discuss what is a good health care for the workers. Gender is a new sector that is very important, because sexism is another issue, that we could have a long interview about here. It exists less than 10 years. We have achieved that in any space of discussion the movement is always represented by one male and one female. That

is an encouragement for woman to come out and speak. Another important sector is the youth sector, where we try to put youth as political beings with specific needs, the goal is to provide youth with the means not to leave the countryside and go to the city. That happens often. Activities: in the struggle of land, the main thing has always been land occupation, but there have also always been other activities. Demonstrations in capital cities, in government bodies relating to land issues. We also work together with other movements, MST is engaged in a broad range of political issues, many fronts where MST is engaged. Within that is the sector of international relations, not only land issue movements in Latin America. The global one is "Vía Campesina". Another sector is culture: we discuss what is the model of popular culture and how to build this model, not just spread regular common widespread bourgois culture. Theater, film and video.

MST Workers / São Paulo

Movimento dos Trabalhadores e Trabalhadoras Sem Teto (Brazil)

MTST emerged out of MST and out of the analysis that the city, the urban, is an important place of struggles.

It started from a perception within MST that they needed to act in the city from their analysis of the movement. They realized in the late 90's that there would be no revolution without action in the cities. They could not disregard cities. Because 85% of the Brazilian population live in cities. Then there was a big discussion within MST, as this idea was evolving, what should bring the activists together: Is it the city and work or is it housing and city? There was a division with two movements then, one called the MTD, the movement of the unemployed workers, and the other one is MTST, the roofless workers movement. MTD is much more of a branch of MST, more closely connected, and although there are people within MTST which are actually part of the MST organisational structure, MTST is a much more independent movement. During this period from 97 to 2001/2 the movement had some very potent occupations, *Anita Origori* with 5000 people in one building. But most important was the occupation of an empty land plot that was owned by Volkswagen. This got the support from German Volkswagen workers. That happened in Lula's first year in his city, where his union is important. That is were his political career began. But back then we were still in a frame of mind that came from MST. In 2004, when I got in contact with MTST, there was a brief occuption in the city of São Paulo, important because of the location. It was evicted by the government fast. By then the movement existed in three states:

> Pernambuco, Rio de Janeiro and São Paulo. But this was a critical moment, the movement almost died because of obstacles we faced in organising people and bringing people together, at least in São Paulo. But it was also a turning point for the movement. In 2005 we had this occupation in another central spot in São Paulo. This is when I think we broke with this frame of mind which was the countryside view on the city. So we then became an actual grassroots city movement. So this occupation in Taho, was very interesting: it brought to the attention of many different movements in São Paulo, especially the south of São Paulo. The occupation was almost in Campo Limpo, a district that is the entrance to the south of the city. So this occupation was very visible for the people in the area. Many people got in touch, went in and went out. New people made contact. That was very interesting. Besides the Hip Hop movement, also the theatre people linked up and got interested. Also from the University of São Paulo people were engaged in that. That was a break with the point of view. This was a real urban occupation. When MST was thinking about MTST, they thought about the periphery of the city. But this occupation brought it in other contexts, building a bridge. This was entirely urban. We dealt with a different set of problems and solutions. We had to deal with the problem of churches in their power in some communities. In one community the catholic priest would be a partner with drug sellers. So, very complex situations regarding the powers in this community.
>
> Fernão / São Paulo

India

Ankur (Delhi)

Ankur differs from others in this work. It can be described as an alternative education program. When I met several members of very different ages, I soon realized that it is hard to understand their principles and goals without having a broader knowledge of India. *Ankur* provides different media labs[1] in working class settlements, set up and grown over the years without official permissions. These settlements are targeted by the municipality and other authorities frequently and are regularly shutdown with force by bulldozers.

> Taking photographs, doing interviews, logging. Logging the particular circumstances of specific places. We use the mobile phones for documentary purposes. The mobile phone is a kind of instrument

1 This is a collaboration with Delhi based *Sarai*. See http://sarai.net.

> that we can take our time as a archive. We go somewhere, take it it to archive. If a locality takes this kind of archiving, so what is this kind of archiving? A person took 400 photographs a year of a particular locality. So, we collect photographs and try to understand what has changed in this year, what does it say about the locality and the person. It could be the meetingplace, the house, anything. Everybody lives intellectual life and wants to archive it. It could be the tea stall, the hair cutting saloon. It is all working class settlements we are talking about.
>
> Anonymous People from Ankur / Delhi

I interviewed a couple of *Ankur* people mainly to get a picture of what mobile media means in their educational program. *Ankur* had been active and present during the brutal shutdown of one of the biggest of such settlements in Delhi.[2] They documented the process and especially the residents' reactions. By approaching mobile media first of all as a way of personal, emotional and intellectual expression, *Ankur* researches with modest means the usage of mobile phones as an amplifier of critical situations. The idea is that young people who learn to express themselves with means such as a mobile phone are in a much better position to express their needs.

Delhi Forum

The Delhi Forum is not about a topic of its own but facilitates the expression of other groups. It takes advantage of the centralization of India, where Delhi is the central power hub.

> I basically participate here in Delhi as a coordinator of the Delhi Forum. This has a simple mandate: to support the people's struggles across the country. Being positioned in Delhi, the capital, means campaign support, legal support. Delhi is the centre as the parliament is here, the constitutional court is here. Delhi has that kind of extra influence over the rest of the country. Delhi Forum has played a key role in terms of lobbying and legal matters. We also do basic support for social movements when they come to Delhi to do demonstrations. We support them, running around for them and organise

2 "Nangla Maanchi was one of the largest settlements on the bank of the river Yamuna, transformed from a fly ash deposit into a lively settlement by its inhabitants. Nangla was demolished in 2006 to make way for the new Riverfront, and some of the settlers have been relocated to Ghevra on the far North of the city." See: http://www.sarai.net/practices/cybermohalla/generative-contexts/locality-labs.

> stuff, also when they are coming to Delhi some of them stay here as a second home. So that they function from here. We coordinate a lot of the national level actions. NAPM, National Forest Workers Forum, National Fish Workers Forum, ... There are different kinds of organisations, many anti-displacement movements, indigenous movements. Women's movement support as well. Delhi Forum has never had to be the key organiser for the women, but does support them too. We also are in a great network about gender and women rights issues.
>
> Vijayan / Delhi

Delhi Solidarity Group

In a similar way like the Delhi Forum, the Delhi Solidarity Group produces sites of protest for organizations that are from others parts of India.

> Here in Delhi I am part of a group called Delhi Solidarity Group. Which tries to support solidarity support in form of organising press conferences, demonstrations, providing advocacy campaign, some publications and do fact finding missions, for movements across the country. Those who call upon us. As part of the Delhi Solidarity Group we also try to organise discussion programs for the members, so that we can ourselves learn and keep in touch with what is happening. In terms of the history or in terms of knowing what positions different actors are taking. So we try to engage on different levels, one is mainly as a solidarity group and also actively being part of some movements. So, for me I am much engaged with "Save Narmada", which is against a big damn being build in a central part of India
>
> Kumar / Delhi

Environmental Support Group (Bangalore)

The ESG is working with local communities about their life conditions in reference to environmental problems. They have an agenda of research to conduct fact finding missions.

> I can define us as a research and campaign group. And we are about 13 years old now. We use the resources that we have to advantage communities. To leverage their issues. We try to critique law and policy in particular. Often times we file investigations. All this helps us to work with learning communities in a far more deeper way. We apply this knowledge in a variety of educational programs

> and we adress school children, college students, teachers, judges. So we work at multiple layers of education. In Bangalore we have been the lead in organising and mobilising a wide range of criticism and responses in terms of technology and environment. Right now we are mobilising people on the issue of the metro destroying the city, its green area. That is one issue which is predominently urban. Also we are a campaign group, holding press conferences. Most of the city's waste is dumped from a bridge. We have been fighting for the last six years to get the landfill shutdown. So there is a serious violation of human rights. We are working with communities about power plants, dams, mining projects and any type of issues they are concerned about. We are also not a funded organisation. We earn our money through research and educational programmes. Which subsidises our campaigns.
>
> Saldanha / Bangalore

National Alliance of People's Movements (India)

This large umbrella organization of many social movements in India tries to bring the many different issues and problems together that in a large country such as India are otherwise dispersed around the country.

> My association is with National Alliance of People's Movement. Which is an umbrella organisation for many movements struggling, mainly for the conservation of the national resources and it is against the corporate and state take over of the national resources, land, water, air. On one hand also on the issues of labour, culture. There is a big thing on the right to work and the right to information. So there is a big movement under this alliance and who work on different issues, mainly those for the marginalised communities
>
> Kumar / Delhi

Sangama, LesBIT (Bangalore)

> We are doing community support for sexual minorities.
>
> Raj and Kannan / Bangalore

Akkai gives a more detailed picture about the practice of this in Bangalore.

> If any violation happens against a sexual minority, if the crisis is there, we take it as a campaign. The Hijras were forced and taken to the police station and they were charged a case against them. Sangana has its own crisis team to negotiate with the police. When we receive a phone call about this, we all gather in front of the

> police station. So then starting to protest. Demanding to release those people. To take down the batton charge against them (this is a beating up charge). Often we do things as a campaign. We also communicate internationally. Taking that issue as a campaign and alerted people. People were very sympathised. That kind of sympathy came from the public. So, then the local people and the political parties demanded to give descend jobs to Hijras. So after that, it was a local change here in Bangalore. But there are more things. We go for the gay lesbian marriage.
>
> Akkai / Bangalore

Hijras are a minority group in India, Pakistan and Bangladesh often called "third gender". They situate themselves neither as males nor females. By doing so they are actively gender-troubling the traditional society and question classic gender roles. They face severe discrimination by the male dominated police forces. *Sangama* also supports sex workers in their struggles for rights, campaigns against forced marriages, and organizes shelter places for young queer, gay, lesbian or intersexual people who are trying to escape from their families and local communities and try to live the way how they feel without harassment.

Japan

No G8 2008 Japan

When the G8 summit in summer 2008 took place on the island of Hokkaidō activists set up a protest infrastructure, an independent media center and organized demonstrations. This followed the examples of the 1999 Seattle WTO protests, which was a starting point for a new protest movement against economic globalization.

> Before last year we already knew that at our town the G8 summit will be held. After getting to know that, we looked to learn at other countries. We made press work to get people to act against the G8 summit. So we set up three or four organisation groups, setting up a camping ground, a media centre and to support NGOs. So our staff seperated into these fields. At that time I was interested in independent media, I wanted get to know the situation, so I joined a group to set up a media centre.
>
> Hosaia / Tokyo

Apart from Yuki Hosaia, Yasudo Yukihiro and Higuchi Takuro are interviewees from this group.

Shiroto No Ran (Tokyo)

Shiroto No Ran is a group from Tokyo: it focuses on neighborhood and community based mutual aid, by operating several second hand stores in an economically declining area of Tokyo. It got famous through humorist activities, such as a candidate campaign for Matsumoto Hajime as Mayor of Tokyo. But this was only a hack: because a mayor candidate is allowed to organize public meetings during campaigning, *Shiroto No Ran* used this loophole to organize open gatherings with music, beer drinking and public debates in places where this is usually strictly forbidden.

> Sometimes we organise demonstrations. Very local based motives. Like a anti-rent demo, or a reclaim the bike demo, or a pro smoking demo, where people come together to smoke together on the street, which is forbidden. My group is called "shiroto no ran" that can be translated to "amateur riot".
>
> Matsumoto / Tokyo

Mexico

Comité Cerezo (Mexico City)

The *Comité Cerezo* is based in Mexico City and is supporting political prisoners in Mexico. The four Cerezo brothers, of whom my interviewee is one, had faced severe repression, because their parents are both living underground in an armed struggle. The brothers of Francisco are imprisoned. I was interested to interview Francisco because he is living under total surveillance. Every move he makes is followed as the authorities think he might contact his parents.

> I am human rights activist, part of the comitee, permantly under threat, like twelve threats already. Involved in different things, like security for political activists
>
> Francisco / Mexico City

There is a trimestrial publication called *Revuelta*, which publishes political articles related directly or indirectly to the cases for seven years now.[3]

3 See http://www.comitecerezo.org.

> First of all we do documentation about the situation of political prisoners, also workshops for human rights groups regarding security.
>
> Francisco / Mexico City

Security for human rights groups as for example good practices of mobile media use and the application of privacy enhancing technologies are amongst these workshops.

LaNeta (Mexico City)

One of the first alternative electronic communication organizations, LaNeta[4] has in many years of work collected experiences about how activists understand and use digital communication means.

> LaNeta is an organization that was born in 1992, it tried since the beginning to offer to other civil society organization communication tools. At that time it was the communication through computers. It was the first intention to enable communications between organization because the organizations were in very difficult period. At that time, the federal government tried to control very tightly not only politically but also financially. So there was a big movement in Mexico from South to North, but communication means were not enough. They needed a special form to communicate fast and cheap. That was our first objective. Then LaNeta built a special service for these organizations, so we have our first server. We started in a very practical sense. Then, we joined APC. In this way we put together different Mexican and organizations from other countries. Especially in terms of social movements, or projects, at that time the internet was not very popular. Then the Zapatisa movement came and LaNeta was a virtual space for that. In some special periods we had more than a 1000 organization that worked through LaNeta. I joined in 1996, in that period LaNeta gave a lot of internet and training workshops, maybe 200, for organizations not only in the cities, but also rural, Oaxaca, Chiapas. With this possibility people could put a lot of information in the internet. At this time, we were the only option for these people. Maybe it is important to say that we were the first server outside universities in 1992. At that time it was only a communication service for other organizations
>
> Olinca / Mexico City

4 See http://www.laneta.apc.org.

APC is the "Association for Progressive Communications", a global umbrella organization with around 50 members in 35 countries.[5] The second interviewee from Laneta explains the range of groups that use Laneta's communication services.

> In general it is a plurality of various groups, from feminist groups, groups that are working with children, indigenous organizations, student movement, human right organizations. And that spectrum includes more leftist groups like some in solidarity with Cuba or related to the Zapatista struggle. The common denominator is that these leftists groups have is social justice. These are the groups that are hosted at LaNeta
>
> Enrique / Mexico City

Pakistan

Concerned Citizen of Pakistan

CCP is mainly comprised of wealthy Lahore citizens that helped organize demonstrations during the lawyer's protests in Pakistan.

> I am the president of this group that we call "society of concerned citizens of Pakistan", or CCP for short. We are two years old. We started in Novembre 2007 after the Chief Justice was sacked. And we are just a group of concerned citizens without any political affiliations who came together to struggle for only one agenda, which was the restauration of the constitution and the rule of law. This was the first time since the 70's that such a movement existed, very long, for two years.
>
> Zaman / Lahore

CCP is in part based on long acquaintances and neighborhood relations.

> For the Lawyer's movement we were part of the concerned citizens of Pakistan. Which were few like-minded people that came together that we knew from our highschool days. We all knew something had to be done, we were not sure what we should do. The CCP were supportive of the lawyer's movements agenda, the restauration of the Chief Justice. The objective was to go after courts, because we knew the dictator had shown his real face by removing the Chief Justice.
>
> Khan / Lahore

5 See http://www.apc.org.

Lawyers' Movement (Pakistan)

The Lawyers' Movement in Pakistan was a reaction to Musharraf's sacking of the Chief Justice in 2007 including other measures to erase an independent jurisdiction (see the case study in chapter 4 for details).

I interviewed two local lawyers that were very active in the protests, Asad Jamal and Mohammad Siddique. Another lawyer was Devin Theriot-Orr. US citizen, he spent the times of the lawyer's movement in Lahore, working for the Human Rights Commission Pakistan and supporting the protests.

> The main activities were getting together, coming out in the streets, e.g. right from the beginning lawyers decided to hold a protest march almost on daily basis, and then it became a weekly affair. I was getting out in most of them. One of the judges, after the imposition of emergency, was served a notice to vacate the house. We spent three nights at his place, raising slogans, showing solidarity. Then we were sending SMS via mobile phones. And at one point of time, when people were started getting arrested, we were holding secret meetings. One was held at my place, and at other friends places. We were following the cases of those who were getting arrested, we were going to the jail standing outside there. On the 10th of decembre, the international human rights day, for the first time after the imposition of emergency, after five weeks, we got together in huge numbers at the high court. We spent two days preparing at different secret places pamphlets, banners, sticks to hold up the banners. The gathering must have been something like 4000 people. Mostly lawyers, but citizens as well. As well as political party workers. Normally lawyers in those days came out with 1500, but that day it was big. Lawyers was one group, obviously the predominant group
>
> Jamal / Lahore

Student Action Committee (Pakistan)

The SAC from LUMS University in Lahore developed an expertise in the use of mobile media and online media during the lawyer's protests. Additionally they were in close contact with oversea students. The SAC worked as an information distribution hub into different media technologies.

Soon after my arrival in Lahore, a bomb went off on an university campus in Islamabad. The consequence was that all universities in the country were closed during most of my stay. As students from private universities, such as Lahore University of Management Sciences (LUMS), had played a crucial role in the protests, this closure made it hard for me to meet students and faculty members. Nonetheless I managed to meet some and conduct two interviews with former students, Abeer Hamid and Ahmed Salemi.

> I was a student that time, we organised rallys, seminars to educate people, long marches, strikes. And a big use of mobile phone. That was the basic thing to mobilise the people.
>
> Abbas, H. / Lahore

Philippines

Linangan Ng Kulturang Pilipino (Manila)

> *Linangan ng kulturang pilipino* is a culture research institute. It deals about raising the level of nationalism of the filippino people, because the American colonialism, or imperialism shall we say, is very much embedded in the filippinos mind: that the americans are good people. But essentially what they did to this country for a hundred years already is really plain exploitation, depravation and repression of the people. That's why we start to educate them about the situation. To eradicate their colonial mentality, their feudal hang ups and also the tendency to be individualists and not to care for their country and countrymen. We do a lot of research and also a lot of education, and also lecture series.
>
> Mina, Minerva, Joan, Julie / Manila

Sinagbayan (Manila)

> *Sinagbayan* means "art for the people", it is cultural mass organisation. It is just one of the cultural organisation here in the Philippines that is working for popularization of progressive arts and culture. So we use different disciplines of art to educate the people, to popularize progressive mass culture. We make theater productions, poetry readings. Most of our productions are into the lifes of the poor people. How we adress the issues of the people. It is always a collective effort, we organise in schools and communities, we organise youth. We conduct workshops, it is free for the communities. To enhance the skills of the youth and serve their communities. And we conduct also education discussions, it is like an alternative progressive education for the youth, where we discuss issues, see how we analyse them in the context of the Philippines struggle. To continue the struggle. We have been working since 2000 and we organise the artists youth students, who study film and also

> artists to network them, so it is a way for them contribute in the social movement and development of the country.
>
> Mina, Minerva, Joan, Julie / Manila

TXTPower (Manila)

SMS is by far the most popular telecommunication method on the South-East Asian archipelago. Due to the extreme precarious economic situation of most of the Philippine people, the costs for SMS play a critical role for social activities. *TXTPower* is a consumer organization that fights for cheap SMS services countrywide. They published a manifesto "Texter's unite! Stand up to the telecom monopolies" in 2001,[6]

> Txtpower started in August 2000, that was a month after the second people power revolt. At that time the mobile companies announced that they are cutting off the free txting allocations. Before, for every 300 peso prepaid card you bought, you get 300 free txt messages. A group of activists formed themselves into TXTpower. The group started a boycott campaign for using txting one day. They wanted to excel some pressure on the phone companies
>
> Eder / Manila

I interviewed Ederic Eder, a member of *TXTPower*, to gather more information about the relevance of SMS for the struggles of activists and social movements. The "second people power revolt" he refers to was an uprising against the corrupt president Estrada, in which according to many partakers mobile media played a crucial role for mobilization. This was in 2001. It is often described as the first SMS revolution, for example by Rheingold (2002).

South Korea

Yongsan Candlelight Media Center (Seoul)

The Seoul based *Yongsan Candlelight Media Center* is set up in the neighborhood of the tragic site of Yongsan, sometimes referred to as the "Yongsan Massacre": On January 20th 2009 a police raid on activists and tenants protesting their eviction from a central building in Seoul caused the death of six people, most of them being

---

6 For the full document, see http://www.tinig.com/v5/v5txt.html (accessed 4 March 2012). It addresses the plans of national telecom operators to raise charges for SMS.

burned. This was followed by large protests, which then again were facing heavy repression. The media center is situated within this neighborhood as a meeting place, social place and media production place.[7]

> We are not an organisation per se. We operate a media centre at the yongsan tragedy site. What we do is basically that we keep an eye on what the police is doing in regard to police violence. We also have performances, concerts in support of the struggle, also we have an exhibition space. We operate a café and also have an oral history project
>
> Two Korean Activists

Spain

Hamlab (Madrid)

Iokese refers to the squat "Patio Maravillas"[8] which was located in central Madrid and which hosted the Hacklab, an autonomous laboratory for technological research. This squat has been evicted on January 5th 2010. The Hacklab was a center of activities for the development and deployment of mobile media tools, such as a SMS emergency tool. Hack labs are spaces of tool and knowledge production. DIY workshops and in general technology centered activities take place here.

> We fight on several levels. On the local level, we help the squat, we help them during the time of the life and when the police comes. We try to make a neighbourhood science place where everybody can come. There is a lot of science experiments in this squatted house. So, that's the local plan. The other projects is trying to use our best force showing the government that we have a new democratic system that makes unnecessary parts of the institutions now. Because we, the citizen, can decide everyday what we want. Citizen represents dignity and a lot of things, so it is a nice word. We made some tools that show people that it is possible to participate in direct actions that can change the reality. Our best practice has been a Xmailer software used in the development of the EU telecom package last year. We gained one more year, because we stopped a vote. We send out more than 400.000 mails a day, giving the media the attention to this problem. The European Parliament stopped the vote and

7 See http://mbout.jinbo.net.

8 See http://defiendelo.patiomaravillas.net

> now they have another year to put the telecom package to work. So, we are not happy with the final result, but it is better now. Another one is the Xvoiper, a software who helps the people to call for free literally to the local phones of the EP members, to explain them that they must vote. We made other tools, like Bluetotem, a software to send bluetooth messages in a demonstration to everyone. We had been collaborating with a lot of European networks to make complaints and stop some problematic laws. Even we had some local actions, like assaulting the building of the Sociedad General de Autores.
>
> Iokese / Madrid

The *Sociedad General de Autores* is the collecting society for artists in Spain. Like the German *GEMA* and the US-american *ASCAP*, this organization is involved with so called "anti-piracy" lobbying and

UK

Consume (London)

Consume is a community based wireless network project in London.[9] Steven James is one of the founding persons behind it. As it provides autonomous wireless media infrastructure, it is one of the few projects that offers insights into mobile media activism in possession of all means of productions. In section 9, Startx and James Stevens are explaining the motivations and activities.

USA

Anti Censorship Activists (New York City)

"Anti Censorship" here stands for the efforts of Nathan Freitas and his colleagues. It does not refer to any specific larger Anti Censorship movement. Nathan and his tech-savvy friends are active in supporting freedom speech in Tibet, Burma and China with sophisticated mobile media technologies that secure these needs.

> There is sort of a subgroup of technologists for work for causes that are similar, that we all want to use the tools of technology for good. Good is sort of loose. For open communication without censorship and with privacy and anonymity when people want it or need it. We

9 Its history is well documented online at http://dek.spc.org/julian/consume/.

are doing a lot of collaborations with different groups, so in support of students for a free Tibet, or the "Free Tibet" campaign, so we work with them and they are activists who perform the action and then the technology folks support it. There is no name, it is not formal, it is sort of this network of people and friends, but if you look at it, we came out of the "Ruckus society". To make a noise. The Ruckus Society came out of Greenpeace and Greenpeace had all this techniques for great protests and they said, wow, we should take these techniques and teach other movements. How can we take our knowledge and share it beyond Greenpeace. And so the Ruckus Society was created. The work I do comes out of that tradition. But the technology aspect is kind of a new thing and is doesn't have a name, it is loose.

Freitas / NYC

Freitas explains the two kinds of activities that this entails:

It is kind of two types of activities. One is support of direct action. So someone is going to do a protest on a mountain or repell off of a building with a banner. Or there is gonna be a large day of action. In that case they do need a technology to support them in communications. So we enable via textmessaging through live video streaming, through satellite communication. We enable communication between the people in the movement and also to the media. And to kind of a global audience. So that's usually around a day of convergence or protest day of action. The second type of activity is more technology development, building tools, training. A longer sighted kind of work. And this is where the project I am working on with Android is focused and also other small projects around this.

Freitas / NYC

The Android project is called "Guardian". Secured communication tools for the Android platform are deployed here, from encrypted chat clients to a port of the anonymity software Tor.[10]

I Witness Video (New York City)

IWV has a very specific orientation in media production.

I am active with a group called "I witness video" which is a group that documents police misconduct. Like Indymedia it started at

10 See https://guardianproject.info/.

> the large convergences, but also like Indymedia it has kind of a community focus.
>
> Goldberg / NYC

The large convergences Goldberg refers to are counter summits during WTO, G8 or G20 meetings.

> "I witness video" is sort of the interesting one in all of this, because they rely on technology more. It is different from Indymedia in that the video collected is primarily for legal defense or to capture a particular bad case of police brutality. But as a strategy more than being journalists. The big thing we do is we go to a demonstration, and we will set up an office and we have people out there with video cameras in teams, but then someone has to go back and log off all the information into a database. There is a lot of technical video work to be done. And they also have a team out there during critical mass in NYC. In NYC you get arrested for critical mass these days. It happens once a month. And also because we have been doing this sorts of work for years we have developed both relationships and expertise that we then use for instance when community groups want to start their own Cop Watch. Many community groups have their own Cop Watch in their own neighbourhood and so we do trainings and exchange information and so on.
>
> Goldberg / NYC

Wireless Seattle

Wireless Seattle has a lot of things in common with London based *Consume*. I have interviewed Matt Westervelt for historical reasons mainly. Seattle is nowadays de facto covered with a municipal wireless network. The urgent need for self-made connectivity is long gone. But still, the experiences made to build an autonomous network infrastructure from below, without any commercial interests, give insightful hints towards hacking media technologies. Matt Westervelt gets extensively quoted in the second part of section 9.

Bibliography

Abbas, A. and S. Jasam (2009). A Ray of Hope: The Case of Lawyers' Movement in Pakistan. In Heinrich Böll Foundation (Ed.), *Pakistan. Reality, Denial and the Complexity of its State*, pp. 140–170.

Abbas, H. / Lahore (2009). Interview with O. Leistert, Lahore, 25.10.2009. Unpublished transcript of audio recording.

Abrahamsen, R. (2003). African Studies and the Postcolonial Challenge. *African Affairs 102*(407), 189–210.

Acquisti, A., S. De Capitani di Vimercati, S. Gritzalis, and C. Lambrinoudakis (Eds.) (2007). *Digital Privacy: Theory, Technologies, and Practices*. New York: Auerbach.

Ahmed, Z. S. (2010). Fighting for the Rule of Law: Civil Resistance and the Lawyers' Movement in Pakistan. *Democratization 17*(3), 492–513.

Akkai / Bangalore (2009). Interview with O. Leistert, Bangalore, 16.10.2009. Unpublished transcript of audio recording.

Alegre / Manila (2009). Interview with O. Leistert, Bangalore, 16.10.2009. Unpublished transcript of audio recording.

Alface / São Paulo (2009). Interview with O. Leistert, São Paulo, 29.7.2009. Unpublished transcript of audio recording.

Allmer, T. (2011). Critical Surveillance Studies in the Information Society. *tripleC 9*(2), 566–592.

Alpha / Mexico City (2009). Interview with O. Leistert, Mexico City, 1.9.2009. Unpublished transcript of audio recording.

Amoore, L. (2011). Data Derivatives. *Theory, Culture & Society 28*(6), 24–43.

Andrejevic, M. (2007). *iSpy: Surveillance and Power in the Interactive Era.* Kansas: University of Minnesota Press.

Anonymous (2010). The Pakistani Lawyers' Movement and the Popular Currency of Judicial Power. *Harvard Law Review 123*, 1705–1726.

Anonymous People from Ankur / Delhi (2009). Interview with O. Leistert, Delhi, 3.10.2009. Unpublished transcript of audio recording.

Anonymous1 / Mexico City (2009). Interview with O. Leistert, Mexico City, 28.8.2009. Unpublished transcript of audio recording.

Anonymous2 / Madrid (2009). Interview with O. Leistert, Madrid, 3.12.2009. Unpublished transcript of audio recording.

Badiou, A. (2009). *L'hypothèse communiste.* Paris: Nouvelles Editions Lignes.

Ball, K. and K. D. Haggerty (2005). Doing surveillance studies. *Surveillance & Society 3*(2/3), 129–137.

Barry, A., T. Osborne, and N. Rose (1996). Introduction. In A. Barry, T. Osborne, and N. Rose (Eds.), *Foucault and Political Reason*, pp. 1–17. Chicago: Chicago University Press.

Beck, U. (1992). *Risk Society: Towards a new Modernity.* London: Sage.

Beniger, J. R. (1986). *The Control Revolution. Technological and Economic Origins of the Information Society.* Cambridge, London: Harvard University Press.

Bentham, J. (1995). *The Panopticon Writings.* London: Verso.

Bey, H. (1991). *The Temporary Autonomous Zone.* New York: Autonomedia.

Bignami, F. (2007). Protecting Privacy against the Police in the European Union: The Data Retention Directive. *Duke Law School Science, Technology and Innovation Research Paper Series, Research Paper No. 13.*

Bispo and Christoph / São Paulo (2009). Interview with O. Leistert, São Paulo, 30.7.2009. Unpublished transcript of audio recording.

Blax / Oaxaca (Mx) (2009). Interview with O. Leistert, Oaxaca, 2.9.2009. Unpublished transcript of audio recording.

Bolognani, M. (2010). Virtual Protest with Tangible Effects? Some Observations on the Media Strategies of the 2007 Pakistani Anti-Emergency Movement. *Contemporary South Asia 18*(4), 401–412.

Bolognani, M. (2011). Rang de Basanti in Pakistan. Elite Student Activism in 2007 Pakistan State of Emergency. In M. Bolognani and S. M. Lyon (Eds.), *Pakistan and its Diaspora: Interdisciplinary Approaches*, pp. 19–42. New York: Palgrave.

Borisov, N. and I. Goldberg (2005). Off-the-Record Communication, or, Why Not to Use PGP. Technical report, CodeCon 2005.

Braman, S. (2006a). *Change of State: Information, Policy, and Power*. Cambridge, MA: MIT Press.

Braman, S. (2006b). Tactical Memory: the Politics of Openness in the Construction of Memory. *First Monday 11*(7).

Bröckling, U. (2007). *Das unternehmerische Selbst: Soziologie einer Subjektivierungsform*. Frankfurt, M.: Suhrkamp.

Bröckling, U., S. Krasmann, and T. Lemke (2010). From Foucault's Lectures at the Collège de France to Studies of Gouvernmentality. An Introduction. In U. Bröckling, S. Krasmann, and T. Lemke (Eds.), *Governmentality. Current Issues and Future Challenges*, pp. 1–33. New York/London: Routledge.

Bublitz, H. (2010a). *Im Beichtstuhl der Medien: Die Produktion des Selbst im öffentlichen Bekenntnis*. Bielefeld: Transcript.

Bublitz, H. (2010b). Täuschen natürlich. Zur Dynamik gesellschaftlicher Automatismen, ihrer Ereignishaftigkeit und strukturbildenden Kraft. In H. Bublitz, R. Marek, C. L. Steinmann, and H. Winkler (Eds.), *Automatismen*, pp. 153–172. München: Fink.

Bunyan, T. (2009). EU Protests: "Troublemakers" database and "travelling violent offenders" (undefined) to be recorded and targeted. *Statewatch Bulletin 19*(3).

Buzan, B., O. Wæver, and J. de Wilde (1998). *Security. A New Framework for Analysis.* Boulder: Lynne Rienner Publishers.

Caminaghi / Campinas (Br) (2009). Interview with O. Leistert, Campinas, 4.8.2009. Unpublished transcript of audio recording.

C.A.S.E. Collective (2006). Critical Approaches to Security in Europe: A Networked Manifesto. *Security Dialogue 37*(4), 443–487.

Castel, R. (1991). From Dangerousness to Risk. In G. Burchell, C. Gordon, and P. Miller (Eds.), *The Foucault Effect: Studies in Governmentality.* Hemel Hampstead: Harvester Wheatsheaf.

Castells, M. (2004). *The Power of Identity. The Information Age: Economy, Society and Culture Vol. II.* (2nd ed.). Oxford: Blackwell.

Castells, M., M. Fernandez-Ardèvol, J. L. Qiu, and A. Sey (2007). *Mobile Communication and Society: A Global Perspective.* Cambridge, MA: MIT Press.

Clarke, R. (1988). Information technology and dataveillance. *Communications of the ACM 31*, 498–512.

Cohen, S. P. (2004). *The Idea of Pakistan.* Washington: Brookings Institution Press.

Collier, S. J. (2009). Topologies of Power. *Theory, Culture & Society 26*(6), 78–108.

Conradi, T. and F. Muhle (2011). Verbinden oder Trennen? Über das schwierige Verhältnis der Akteur-Netzwerk-Theorie zur Kritik. In T. Conradi, H. Derwanz, and F. Muhle (Eds.), *Strukturentstehung durch Verflechtung. Akteur-Netzwerk-Theorie(n) und Automatismen*, pp. 313–333. München: Fink.

Cox, L. and A. G. Nilsen (2007). Social Movements Research and the 'Movement of Movements': Studying Resistance to Neoliberal Globalisation. *Sociology Compass 1*(2), 424–442.

Coyle, D. (2005, March). Africa: The Impact of Mobile Phones. The Vodafone Policy Paper Series, 2.

Danezis, G. and S. Gürses (2010). A Critical Review of 10 Years of Privacy Technology. In *Proceedings of Surveillance Cultures: A Global Surveillance Society?* http://homes.esat.kuleuven.be/~sguerses/papers/DanezisGuersesSurveillancePets2010.pdf (accessed 23 April 2012).

Danezis, G. and B. Wittneben (2006). The Economics of Mass Surveillance - and the Questionable Value of Anonymous Communications. In *Proceedings of the 5th Workshop on The Economics of Information Security (WEIS 2006).*

de Souza e Silva, A., D. M. Sutko, F. A. Salis, and C. de Souza e Silva (2011). Mobile phone appropriation in the favelas of Rio de Janeiro, Brazil. *New Media & Society 13*(3), 411–426.

Dean, M. (2010). *Governmentality: Power and Rule in Modern Society*. London: Sage. Second edition.

Death, C. (2010). Counter-conducts: A Foucauldian Analytics of Protest. *Social Movement Studies 9*(3), 235–251.

Death, C. (2011a). Africa, Governmentality, and the International: Reflections on Agency and the Climate Change Negotiations. In *BISA-Africa and International Studies ESRC seminar series nr. 4: African agency: Implications for International Relations theory.*

Death, C. (2011b). Counter-conducts in South Africa: Power, Government and Dissent at the World Summit. *Globalizations 8*(4), 425–438.

Deleuze, G. (1992). Postscript on the Societies of Control. *October 59*, 3–7. First published 1990.

Della Porta, D. and M. Diani (2006). *Social Movements: an Introduction*. Oxford: Blackwell. Second edition.

Della Porta, D., H. Kriesi, and D. Rucht (Eds.) (2009). *Social Movements in a Globalizing World.* New York: Macmillan. Second expanded edition.

Dencik, L. (2012). *Media and Global Civil Society*. Basingstoke/New York: Palgrave Macmillan.

Diffie, W. and S. Landau (1998). *Privacy on the Line. The Politics of Wiretapping and Encryption.* Cambridge, MA: MIT Press.

Dominguez / San Diego (2009). Interview with O. Leistert, Madrid, 4.12.2009. Unpublished transcript of audio recording.

Donner, J. (2009). Blurring Livelihoods and Lives. The Social Uses of Mobile Phones and Socioeconomic Development. *Innovations 4*(1), 91–102.

Dunbar-Hester, C. (2009). 'Free the Spectrum!' - Activist Encounters with Old and New Media Technology. *New Media & Society 11*(1&2), 221–240.

Eder / Manila (2009). Interview with O. Leistert, Manila, 3.10.2009. Unpublished transcript of audio recording.

Engel, T. (2007). *Sichere Nachrichtenübermittlung mit Off-the-Record-Messaging*. Diplomarbeit, Technische Fachhochschule Berlin.

Enrique / Mexico City (2009). Interview with O. Leistert, Mexico City, 28.9.2009. Unpublished transcript of audio recording.

European Commission (2006). European Union Directive 2006/24/EC of March 2006 on the retention of data generated or processed in connection with the provision of publicly available electronic communications services or of public communication networks and amending Directive 2002/58/EC. *Official Journal of the European Union L 105*, 56–63.

Fagan, K. (2011). Man shot to death by BART officer identified. San Francisco Chronicle. July 8. http://www.sfgate.com/cgi-bin/article.cgi?f=/c/a/2011/07/07/BA9U1K7O2C.DTL (accessed 4 March 2012).

Fenton, N. and V. Barassi (2011). Alternative media and social networking sites: The politics of individuation and political participation. *The Communication Review 14*(3), 179–196.

Ferguson, N. and B. Schneier (2003). *Practical Cryptography*. Indianapolis: Wiley.

Fernão / São Paulo (2009). Interview with O. Leistert, São Paulo, 2.8.2009. Unpublished transcript of audio recording.

Fernandez, L. A. and L. Huey (2009). Is Resistance Futile? Thoughts on Resisting Surveillance. *Surveillance & Society 6*(3), 198–202.

Foucault, M. (1977). *Discipline and Punish: Birth of the Prison.* New York: Pantheon Books. First published 1976.

Foucault, M. (1997a). Security, Territory, and Population. In P. Rabinow (Ed.), *Ethics: Subjectivity and Truth*, Translated R. Hurley et al., pp. 67–71. New York: The New Press.

Foucault, M. (1997b). Subjectivity and Truth. In P. Rabinow (Ed.), *Ethics: Subjectivity and Truth*, Translated R. Hurley et al., pp. 87–92. New York: The New Press.

Foucault, M. (1997c). The Birth of Biopolitics. In P. Rabinow (Ed.), *Ethics: Subjectivity and Truth*, Translated R. Hurley et al., pp. 73–79. New York: The New Press.

Foucault, M. (2000a). Lives of Infamous Men. In J. D. Faubion (Ed.), *Power*, Translated R. Hurley et al., pp. 155–175. New York: The New Press.

Foucault, M. (2000b). Questions of Method. In J. D. Faubion (Ed.), *Power*, Translated R. Hurley et al., pp. 223–238. New York: The New Press. First published 1980.

Foucault, M. (2000c). Sex, Power, and the Politics of Identity. In P. Rabinow (Ed.), *Ethics*, pp. 163–174. London/New York: Penguin. First published 1984.

Foucault, M. (2000d). Space, Knowledge, and Power. In J. D. Faubion (Ed.), *Power*, Translated R. Hurley et al., pp. 349–364. New York: The New Press. First published 1982.

Foucault, M. (2000e). The Risks of Society. In J. D. Faubion (Ed.), *Power*, Translated R. Hurley et al., pp. 365–381. New York: The New Press. First published 1983.

Foucault, M. (2000f). The Subject and Power. In J. D. Faubion (Ed.), *Power*, Translated R. Hurley et al., pp. 326–348. New York: The New Press. First published 1982.

Foucault, M. (2000g). "Omnes et singulatim": Toward a Critique of Political Reason. In J. D. Faubion (Ed.), *Power*, Translated R. Hurley et al., pp. 298–325. New York: The New Press. These are the two Tenner lectures held October 1979.

Foucault, M. (2003). *Society Must Be Defended: Lectures at the Collège de France, 1975 - 1976.* Translated by David Macey. New York: Picador. First published 1997.

Foucault, M. (2007). *Security, Territory, Population. Lectures at the Collège de France, 1977 - 1978.* Translated by Graham Burchell. New York: Picador. First published 2004.

Foucault, M. (2008). *The Birth of Biopolitics. Lectures at the Collège de France, 1978 - 1979.* Translated by Graham Burchell. New York: Picador. First published 2004.

Francisco / Mexico City (2009). Interview with O. Leistert, Mexico City, 4.9.2009. Unpublished transcript of audio recording.

Freddie / Hong Kong (2009). Interview with O. Leistert, Tokyo, 25.9.2009. Unpublished transcript of audio recording.

Freitas / NYC (2009). Interview with O. Leistert, New York City, 12.8.2009. Unpublished transcript of audio recording.

Fuller, M. (2007). *Media Ecologies. Materialist Energies in Art and Technoculture.* Cambridge, MA: MIT Press.

Fuller, M. (2008). *Software Studies. A Lexicon.* Cambridge, MA: MIT Press.

Gallagher, R. and R. Syal (2011a). Met Police Using Surveillance System to Monitor Mobile Phones. The Guardian. October 30. http://www.guardian.co.uk/uk/2011/oct/30/metropolitan-police-mobile-phone-surveillance (accessed 20 March 2012).

Gallagher, R. and R. Syal (2011b). Police Buy Software to Map Suspects' Digital Movements. The Guardian. May 11. http://www.guardian.co.uk/uk/2011/may/11/police-software-maps-digital-movements (accessed 20 March 2012).

Galloway, A. R. (2004). *Protocol: How Control Exists After Decentralization*. Cambridge, Mass: MIT Press.

Galperin, E. (2011). Want Public Safety? Don't Disable Cell Phones. EFF. August 19. https://www.eff.org/deeplinks/2011/08/want-public-safety-dont-disable-cell-phones (accessed 4 March 2012).

Geirbo, H. C. and P. Helmersen (2008). Turning Threats into Opportunities – the Social Dynamics of Missed Calls. *Telektronikk* (2), 77–83.

Goggin, G. (2006). *Cell Phone Culture. Mobile Technology in Everyday Life*. London/New York: Routledge.

Goldberg / NYC (2009). Interview with O. Leistert, New York City, 7.8.2009. Unpublished transcript of audio recording.

Grassmuck, V. (2002). *Freie Software. Zwischen Privat- und Gemeineigentum*. Bonn: Bundeszentrale für politische Bildung.

Habermas, J. (1987). *Theory of Communicative Action*. Cambridge: Polity.

Haggerty, K. and R. Ericson (2000). The Surveillant Assemblage. *British Journal of Sociology 51*, 605–622.

Haraway, D. J. (1985). A Manifesto for Cyborgs: Science, Technology and Socialist Feminism in the 1980s. *Socialist Review 80*, 65–108.

Haraway, D. J. (1997). *Modest Witness@Second Millennium. FemaleMan Meets OncoMouse: Feminism and Technoscience*. New York: Routledge.

Hauck, M. (2012). Niedersachsen jagt Verbrecher mittels eines privaten Dienstleisters. Süddeutsche Zeitung. February 27. http://www.sueddeutsche.de/digital/ueberwachung-per-stiller-sms-niedersachsen-jagt-verbrecher-mittels-eines-privaten-dienstleisters-1.129500 (accessed 4 March 2012).

Haunss, S. (2005). Geschichte und Perspektiven sozialer Bewegungen. In G. O. Hüttner and N. Schepers (Eds.), *Vorwärts und viel vergessen - Beiträge zur Geschichte und Geschichtsschreibung neuer sozialer Bewegungen*, pp. 27–42. Neu-Ulm: AG Spak.

Hellmann, K.-U. (2002). Paradigmen der Bewegungsforschung. In K.-U. Hellmann and R. Koopmans (Eds.), *Paradigmen der Bewegungsforschung*, pp. 9–32. Wiesbaden: VS Verlag für Sozialwissenschaften.

Herkenrath, M. (2009). *¡Hasta la victoria siempre! Linksprogressive soziale Bewegungen und ihre transnationalen Koalitionen.* Philosophische Fakultät der Universität Zürich: Habilitationsschrift.

Hijuchi / Tokyo (2009). Interview with O. Leistert, Tokyo, 26.9.2009. Unpublished transcript of audio recording.

Holland, N. (2002). Truth as Force. Michel Foucault on Religion, State Power, and the Law. *Journal of Law and Religion 18*(1), 79–97.

Hosaia / Tokyo (2009). Interview with O. Leistert, Tokyo, 25.9.2009. Unpublished transcript of audio recording.

Housley, R. and T. Polk (2001). *Planning for PKI. Best Practices Guides for Deploying a Public Key Infrastructure.* New York: Wiley.

Howard, P. N., S. D. Agarwal, and M. M. Hussain (2011). When Do States Disconnect Their Digital Networks? Regime Responses to the Political Uses of Social Media. *The Communication Review 14*(3), 216–232.

Hoy, D. C. (2004). *Critical Resistance. From Poststructuralism to Post-Critique*. Cambridge, MA: MIT Press.

Huxley, A. (1932). *Brave New World.* London: Chatto & Windus.

Illouz, E. (2008). *Saving the Modern Soul: Therapy, Emotions, and the Culture of Self-Help.* University of California Press.

Innis, H. A. (2007). *Empire and Communications.* Toronto: Dundurn Press. First published 1950.

Iokese / Madrid (2009). Interview with O. Leistert, Madrid, 4.12.2009. Unpublished transcript of audio recording.

Iqbal, N. (2009). A View from Pakistan. *Human Rights 36*(1), 21–23.

Jamal / Lahore (2009). Interview with O. Leistert, Lahore, 28.10.2009. Unpublished transcript of audio recording.

Jones, O. B. (2005). *Pakistan. Eye of the Storm*. London/New York: Pinguin. Original published in 2002 by Yale University Press.

Juris, J. S. (2008). *Networking Futures. The Movements Against Corporate Globalisation*. Durham/London: Duke University Press.

Kahn Gillmor / NYC (2009). Interview with O. Leistert, New York City, 11.8.2009. Unpublished transcript of audio recording.

Kalpagam, U. (2000). Colonial Governmentality and the 'Economy'. *Economy & Society 29*(3), 418–438.

Khan / Lahore (2009). Interview with O. Leistert, Lahore, 27.10.2009. Unpublished transcript of audio recording.

Krasmann, S. and M. Volkmer (2007). Einleitung. In S. Krasmann and M. Volkmer (Eds.), *Michel Foucaults "Geschichte der Gouvernementalität" in den Sozialwissenschaften. Internationale Beiträge*, pp. 7–20. Bielefeld: Transcript.

Kumar / Delhi (2009). Interview with O. Leistert, Delhi, 17.10.2009. Unpublished transcript of audio recording.

Landau, S. (2010). *Surveillance or Security? The Risks Posed by New Wiretapping Technologies*. Cambridge, MA: MIT Press.

Le Bodic, G. (2005). *Mobile Messaging. Technologies and Services. SMS, EMS, and MMS*. West Sussex: Wiley.

Legume / São Paulo (2009). Interview with O. Leistert, São Paulo, 29.7.2009. Unpublished transcript of audio recording.

Leistert, O. (2008). Data Retention in the European Union: When a Call Returns. *International Journal of Communication 2*, 925–935.

Leistert, O. (2010). Individuation, Nachbarschaft und Protokoll. In M. Bierwirth, O. Leistert, and R. Wieser (Eds.), *Ungeplante Strukturen. Tausch und Zirkulation*, pp. 33–46. München: Fink.

Leistert, O. (2012). The iPhone's Failure: Protests and Resistances. In P. Snickars and P. Vonderau (Eds.), *Moving Data: The iPhone and the Future of Media*. New York: Columbia UP.

Lemke, T. (2002). Foucault, Governmentality, and Critique. *Rethinking Marxism 14*(3).

Lemke, T. (2007). An Indigestible Meal? Foucault, Governmentality and State Theory. *Distinktion: Scandinavian Journal of Social Theory* (15).

Lemons, K. (2008). Gouvernementalität, Religion und soziale Sicherheit im postkolonialen Indien. In P. Purtschert, K. Meyer, and Y. Winter (Eds.), *Gouvernementalität und Sicherheit. Zeitdiagnostische Beiträge im Anschluss an Foucault*. Bielefeld: Transcript.

Lessig, L. (1999). *Code*. New York: Basic Books.

Lianos, M. (2003). Social Control after Foucault. *Surveillance & Society 3*(1), 412–430.

Ling, R. (2004). *The Mobile Connection. The Cell Phone's Impact On Society.* San Francisco: Morgan Kaufmann.

Lyon, D. (2001). *Surveillance Society. Monitoring Everyday Life*. Buckingham/Philadelphia: Open University Press.

Lyon, D. (2003). Surveillance as Social Sorting: Computer Codes and Mobile Bodies. In D. Lyon (Ed.), *Surveillance as Social Sorting. Privacy, Risk, and Digital Discrimination*, pp. 13–30. New York: Routledge.

Lyon, S. M. (2010). Cell Phones and People Power: The Impact of Information Technologies on Politics in Pakistan. Unpublished Manuscript.

M / São Paulo (2009). Interview with O. Leistert, São Paulo, 29.7.2009. Unpublished transcript of audio recording.

Malik, M. A. (2008). *The Pakistan Lawyer's Movement. An Unfinished Agenda.* Karachi: Pakistan Law House.

ManasaSarovara / Bangalore (2009). Interview with O. Leistert, Bangalore, 14.10.2009. Unpublished transcript of audio recording.

Manovich, L. (2001). *The Language of New Media.* Cambridge, MA: MIT Press.

Marks, P. (2005). Imagining Surveillance: Utopian Visions and Surveillance Studies. *Surveillance & Society 3*(2/3), 222–239.

Marquis, G. (2003). Private Security and Surveillance. In D. Lyon (Ed.), *Surveillance as Social Sorting. Privacy, Risk and Digital Discrimination*, pp. 227–248. London/New York: Routledge.

Marx, G. T. (1985). The Surveillance Society: The Threat of 1984-style Techniques. *The Futurist 6*, 21–26.

Massumi, B. (2005). Fear (the spectrum said). *Positions. East Asia Cultures Critique 13*(1), 31–48.

Massumi, B. (2009). National Enterprise Emergency. *Theory, Culture & Society 26*(6), 153–185.

Matsumoto / Tokyo (2009). Interview with O. Leistert, Tokyo, 26.9.2009. Unpublished transcript of audio recording.

Mattelart, A. (2010). *The Globalization of Surveillance.* Cambridge: Polity.

May, T. (1994). *The Political Philosophy of Poststructuralist Anarchism.* University Park: Penn State University Press.

McLuhan, M. and T. Gordon (2003). *Understanding Media. The Extensions of Man.* Corte Madera, CA: Gingko Press. First published 1964.

Meister, A. (2008). Vorratsdatenspeicherung und gesellschaftliche Kontroverse. Bachelor thesis. Humboldt-Universität zu Berlin.

Menezes, A. J., P. C. V. Oorschot, S. A. Vanstone, and R. L. Rivest (1997). *Handbook of Applied Cryptography.* Boca Raton, FL: CRC Press.

Meuser, M. and U. Nagel (2005). Expertinneninterviews - vielfach erprobt, wenig bedacht. In A. Bogner, B. Littig, and W. Menz (Eds.), *Das Experteninterview. Theorie, Methode, Anwendung.*, pp. 71–94. Wiesbaden: VS Verlag für Sozialwissenschaften. Second edition.

Miller, P. and N. Rose (2008). *Governing the Present: Administering Economic, Social and Personal Life*. Cambridge: Polity Press.

Mina, Minerva, Joan, Julie / Manila (2009). Interview with O. Leistert, Manila, 3.10.2009. Unpublished transcript of audio recording.

Mitnick, K. D. (2002). *The Art of Deception. Controlling the Human Element of Security*. Indianapolis: Wiley.

Mitnick, K. D. (2005). *The Art of Intrusion*. Indianapolis: Wiley.

Mohammad Azhar Siddique / Lahore (2009). Interview with O. Leistert, Lahore, 21.10.2009. Unpublished transcript of audio recording.

Montgomery, N. (2010). Beyond Civil Disobedience and Counter-hegemony: Legitimacy, Strategy and Tactics in the 2010 Anti-Olympics Movement. In F. Heßdörfer, A. Pabst, and P. Ullrich (Eds.), *Prevent and Tame. Protest under (Self-) Control*, pp. 99–118. Berlin: Karl Dietz Verlag.

MST Workers / São Paulo (2009). Interview with O. Leistert, São Paulo, 31.7.2009. Unpublished transcript of audio recording.

Munir, D. (2009). Struggling for the Rule of Law: The Lawyers' Movement. *Middle East Report 39*(251). http://www.merip.org/mer/mer251/struggling-rule-law (accessed 23 April 2012).

Negri, T. and M. Hardt (2000). *Empire*. London: Harvard University Press.

Non Collective / Manila (2009). Interview with O. Leistert, Manila, 2.10.2009. Unpublished transcript of audio recording.

Norris, C. and G. Armstrong (1999). *The Maximum Surveillance Society: The Rise of CCTV*. Oxford: Berg.

Odysseos, L. (2011). Governing Dissent in the Central Kalahari Game Reserve: 'Development', Governmentality, and Subjectification amongst Botswana's Bushmen. *Globalizations 8*(4), 439–455.

Olinca / Mexico City (2009). Interview with O. Leistert, Mexico City, 1.9.2009. Unpublished transcript of audio recording.

Opitz, S. (2010). Government Unlimited. The Security Dispositif of Illiberal Governmentality. In U. Bröckling, S. Krasmann, and T. Lemke (Eds.), *Governmentality. Current Issues and Future Challenges*, pp. 93–114. New York/London: Routledge.

Orwell, G. (1949). *Nineteen Eighty-Four*. London: Secker & Warburg.

O´Harrow, R. (2006). *No Place to Hide*. New York: Free Press.

Pariser, E. (2011). *The Filter Bubble. What the Internet is Hiding from You*. London: Penguin.

Pertierra, R. (2006). *Transforming Technologies: Altered Selves*. Manila: De La Salle University Press.

Pertierra, R. (2008). The new media and heterotopic technologies in the philippines. *Human Technology 4(2)*, 169–185.

Pertierra, R., E. Ugarte, A. Pingol, J. Hernandez, and N. L. Dacanay (2002). *Txting Selves. Cellphones and Philippine Modernity*. Manila: De La Salle University Press.

Plant, S. (2002). *On the Mobile. The Effects of Mobile Telephones on Social and Individual Life*. Published by Motorola. http://classes.dma.ucla.edu/Winter03/104/docs/splant.pdf (accessed 23 April 2012).

Priest, J. and J. Stevens (2004). Trip the Loop, Make your Switch, Consume the Net. In G. Cox, J. Krysa, A. Lewin, and M. Miless (Eds.), *Economising Culture*, pp. 121–132. Brooklyn: Autonomedia.

Rafael, V. L. (2003). The Cell Phone and the Crowd: Messianic Politics in the Contemporary Philippines. *Public Culture 15*(3), 399–425.

Raj and Kannan / Bangalore (2009). Interview with O. Leistert, Bangalore, 15.10.2009. Unpublished transcript of audio recording.

Rao, A. and P. Dasgupta (2010). Rural Livelihoods and Income Enhancement in the "New" Economy. 4th ACM Workshop on Networked Systems for Developing Regions. June 15th, 2010.

Reyes / Manila (2009). Interview with O. Leistert, Manila, 2.10.2009. Unpublished transcript of audio recording.

Rheingold, H. (2002). *Smart Mobs: The Next Social Revolution*. Cambridge, MA: Perseus Publishing.

Rose, N. (1996). Governing 'Advanced' Liberal Democracies. In A. Barry, T. Osborne, and N. Rose (Eds.), *Foucault and political reason*, pp. 37–64. Chicago: Chicago University Press.

Royal Academy of Engineering (2007). *Dilemmas of Privacy and Surveillance. Challenges of Technological Change.* http://www.raeng.org.uk/news/publications/list/reports/dilemmas_of_privacy_and_surveillance_report.pdf (accessed 23 April 2012).

Röhle, T. (2011). Strategien ohne Strategen. Intentionalität als ‚Strukturentstehung durch Verflechtung'? In T. Conradi, H. Derwanz, and F. Muhle (Eds.), *Strukturentstehung durch Verflechtung. Akteur-Netzwerk-Theorie(n) und Automatismen*, pp. 173–192. München: Fink.

Saar, M. (2010). Relocating the Modern State. Governmentality and the History of Political Ideas. In U. Bröckling, S. Krasmann, and T. Lemke (Eds.), *Governmentality. Current Issues and Future Challenges*, pp. 34–55. New York/London: Routledge.

Saldanha / Bangalore (2009). Interview with O. Leistert, Bangalore, 17.10.2009. Unpublished transcript of audio recording.

Salemi / Lahore (2009). Interview with O. Leistert, Lahore, 25.10.2009. Unpublished transcript of audio recording.

Samatas, M. (2005). Studying Surveillance in Greece: Methodological and Other Problems Related to an Authoritarian Surveillance Culture. *Surveillance & Society 3*(2/3), 181–197.

Schneier, B. (1996). *Applied Cryptography, Second Edition. Protocols, Algorithms, and Source Code.* Indianapolis: Wiley.

Scott, D. (1995). Colonial Governmentality. *Social Text* (43), 191–220.

Seier, A. (2011). Un/Verträglichkeiten: Latours Agenturen und Foucaults Dispositive. In T. Conradi, H. Derwanz, and F. Muhle (Eds.), *Strukturentstehung durch Verflechtung. Akteur-Netzwerk-Theorie(n) und Automatismen*, pp. 151–172. München: Fink.

Shaheen, M. A. (2008). Use of Social Networks and Information Seeking Behavior of Students During Political Crises in Pakistan: A Case Study. *The International Information & Library Review 40*, 142–147.

Siegert, B. (1993). *Relais. Geschicke der Literatur als Epoche der Post 1751 - 1913*. Berlin: Brinkmann & Bose.

Sigley, G. (2006). Chinese Governmentalities: Government, Governance and the Socialist Market Economy. *Economy & Society 35*(4), 487–508.

Sinha, C. (2005). Effect of Mobile Telephony on Empowering Rural Communities in Developing Countries. International Research Foundation for Development. Conference on Digital Divide, Global Development and the Information Society. November 14-16.

Soghoian, C. and S. Stamm (2011). Certified Lies: Detecting and Defeating Government Interception Attacks Against SSL. International Financial Cryptography Association. Financial Cryptography and Data Security '11.

Startx / London (2009). Interview with O. Leistert, London, 6.4.2009. Unpublished transcript of audio recording.

Stevens / London (2009). Interview with O. Leistert, London, 6.4.2009. Unpublished transcript of audio recording.

Stoler, A. (1995). *Race and the Education of Desire: Foucault's History of Sexuality and the Colonial Order of Things.* Durham: Duke University Press.

Strandburg, K. J. (2007). Surveillance of Emergent Associations: Freedom of Association in a Network Society. In A. Acquisti, S. D. C. di Vimercati, S. Gritzalis, and C. Lambrinoudakis (Eds.), *Digital Privacy: Theory, Technologies, and Practices*, pp. 435–458. New York: Auerbach.

Swapan / Dhaka (2009). Interview with O. Leistert, Tokyo 22.9.2009. Unpublished transcript of audio recording.

Tenhunen, S. (2008). Mobile Technology in the Village: ICTs, Culture, and Social Logistics in India. *Journal of the Royal Anthropological Institute 14*(3), 515–534.

Terranova, T. (2004). *Network Culture. Politics for the Information Age.* London/Ann Arbour: Pluto Press.

Theriot-Orr / Seattle (2009). Interview with O. Leistert, Seattle, 24.8.2009. Unpublished transcript of audio recording.

Tomitsch, M., F. Sturm, M. Konzett, A. Bolin, I. Wagner, and T. Grechenig (2010). Stories from the Field: Mobile Phone Usage and its Impact on People's Lives in East Africa. In *Proceedings of the International Conference on Information and Communication Technologies and Development (ICTD'10).*

Two Korean Activists (2009). Interview with O. Leistert, Tokyo, 25.9.2009. Unpublished transcript of audio recording.

Töpfer, E. (2009). Germany: A Network Being Networked: the Federal Criminal Police Office Databases and the Surveillance of "Troublemakers". *Statewatch Bulletin 19*(3).

United Nations Development Programme (2009). *Human Development Report 2009.* Basingstoke/New York: Palgrave Macmillan.

Vijayan / Delhi (2009). Interview with O. Leistert, Delhi, 3.11.2009. Unpublished transcript of audio recording.

Vora, A. (2007). Pakistan's Students Push for Democracy. *New Internationalist Magazine* (407). http://www.newint.org/features/special/2007/12/14/pakistan-democracy/ (accessed 23 April 2012).

Völker, C. (2010). *Mobile Medien. Zur Genealogie des Mobilfunks und zur Ideengeschichte von Virtualität.* Bielefeld: Transcript.

Wallis, C. (2011). Mobile Phones without Guarantees: The Promises of Technology and the Contingencies of Culture. *New Media & Society 13*(3), 471–485.

Warner, J. (2005). The Right to Oblivion: Data Retention from Canada to Europe in Three Backward Steps. *University of Ottawa Law & Technology Journal 2*, 77–104.

Westervelt / Seattle (2009). Interview with O. Leistert, Seattle, 24.8.2009. Unpublished transcript of audio recording.

Winkler, H. (2004). *Diskursökonomie. Versuch über die innere Ökonomie der Medien.* Frankfurt am Main: Suhrkamp.

Winkler, H. and B. Tischleder (2001). Portable Media. *Ästhetik & Kommunikation* (112), 97–104.

Winner, L. (1977). *Autonomous Technology: Technics out of Control as a Theme in Human Thought.* Cambridge, MA: MIT Press.

Wood, D. M. (2003). Editorial. Foucault and Panopticism Revisited. *Surveillance & Society 1*(3), 234–239.

Wood, D. M. (2007). Surveillance in Japan: A Critical Introduction. *Urban Studies 44*(3), 551–568.

Wood, D. M. (2009). The 'Surveillance Society': Questions of History, Place and Culture. *European Journal of Criminology 6*, 179–194.

WSMC (2011). EFF Austin. Statement on San Francisco BART Cellphone Service Shutdown. August 14. http://effaustin.org/2011/08/statement-on-san-francisco-bart-cellphone-service-shutdown (accessed 4 March 2012).

Yasuda / Tokyo (2009). Interview with O. Leistert, Tokyo, 19.9.2009. Unpublished transcript of audio recording.

Yusuf, H. (2009). Old and New Media: Converging During the Pakistan Emergency (March 2007- February 2008). MIT Center for Future Civic Media. http://civic.mit.edu/sites/civic.mit.edu/files/Old%20and%20New%20Media%20Pakistan%20Emergency.pdf (accessed 20 April 2012).

Zaman / Lahore (2009). Interview with O. Leistert, Lahore, 27.10.2009. Unpublished transcript of audio recording.

Zimmermann, P. (1995). Author's preface to the book: PGP Source Code and Internals. http://www.philzimmermann.com/EN/essays/BookPreface.html (accessed 24 April 2012). Originally published by MIT Press, no longer in print.

Glossary and Abbreviations

ACLU American Civil Liberties Union. A non-profit US organization that engages in the protection of citizen's rights, like the freedom of speech.

Authentication in technical terms a mechanism that ensures the authenticity of the message sender on a mathematical base.

CAE Critical Art Ensemble. A group of artists/activists that try to shift the discourse to question technology and society. CAE uses digital technologies and biotechnology as practical ways of disturbing and intervening into public discourse.

CCP Concerned Citizen of Pakistan. A group based in Lahore, mostly with a wealthy background, that was formed to support the demands of the lawyers.

CMI Centro di Midia Independente. Indymedia from Brazil..

Confidentiality In digital communication technology, confidentiality means that a message can not be read by an unintended party.

DDoS Distributed denial of service. From many distributed networked computers are huge amounts of ordinary HTTP requests sent to one server; more than the server can handle. This is leading to a denial of service at the targeted server, rendering it inoperative.

Deniability To be able to deny a message in digital communication technology translates into the guaranteed singularity of each encrypted message: even if one message is deciphered by third parties, past and future ones are not.

DoD US Department of Defense.

ECD Electronic Civil Disobedience. An attempt to transpose civil disobedience into the digital realm. A website is the place of a virtual sit-in, which is manifested by reoccurring reloads of the page from many partakers. Special software has been developed to reload the site automatically, thus it is sometimes regarded as a distributed denial of service attack.

EDT Electronic Disturbance Theater.

EFF Electronic Frontier Foundation. A non-profit organization that engages in the protection of citizen rights in the digital domain.

ESG Environmental Support Group.

HTTPS HTTP secured with SSL.

IMEI International Mobile Station Equipment Identity is a 15-digits long, unique number, which makes it possible to identify any GSM or UMTS device.

IMSI International Mobile Subscriber Identity is a unique number which identifies all SIM-cards.

Integrity in technical terms is a mechanism that ensures on a mathematical base the integrity of the message.

LAC Location Area Cell is the current cell a cell phone is logged in.

LEA Law Enforcement Agencies: an umbrella term for all governmental and non-official governmental bodies from police to secret service.

LUMS Lahore University of Management Sciences is a prestigious private University in Lahore, Pakistan.

MCC Mobile Country Code. A code defined by the ITU that allocates a numerical code for the country of the telecom provider of mobile phones. It is a part of the IMSI.

MG Militant Group. A phantom-like group, or maybe more than one, of radical activists inhibiting force in their actions against the German state and companies. No attacks on people are known.

MNC Mobile Network Code. A code to uniquely identify a mobile phone operator/carrier using one of the common mobile communications standards such as GSM.

MPL Movimento Passe Livre.

MSC Mobile Switching Center. The primary service delivery node for GSM and CDMA, responsible for routing voice calls and SMS as well as other services (such as conference calls, FAX and circuit switched data). The MSC sets up and releases the end-to-end connection, handles mobility and handover requirements during the call and takes care of charging and real time pre-paid account monitoring.

MST Movimento dos Trabalhadores Rurais Sem Terra.

MTST Movimento dos Trabalhadores e Trabalhadoras Sem Teto.

NACL National Action Committee of Lawyers. The main decision body of the lawyers' movement in Pakistan.

NAPM National Alliance of People's Movements.

OpenPGP Encryption and key management application with a wide range of encryption, decryption, authentication functionalities. Often used to encrypt email.

OTR Off-the-Record Messaging is an encrypted messaging style that offers perfect forward secrecy (PFS) and thus deniability.

PFS Perfect forward secrecy.

PGP Encryption application that historically made encryption available for common users.

PNR Passenger Name Records are the data sets that airlines collect about their customers. Recent legislation in the European Union provides the retention of this data at the US Department of Homeland Security with only very weak anonymization and for more than a decade, including PNR of flights within the EU.

SAC The Student Action Committee was initiated at the LUMS in Lahore during martial law.

SMTP Simple Mail Transfer Protocol. The protocol to send emails.

www.ingramcontent.com/pod-product-compliance
Ingram Content Group UK Ltd.
Pitfield, Milton Keynes, MK11 3LW, UK
UKHW021831100926
8556IPUK00004B/278

9 783631 643136